9/25/96

D0425619

Color and symmetry

Wiley monographs in crystallography

Polymorphism and polytypism in crystals
 Ajit Ram Verma and P. Krishna

Molecular crystals
 J. L. Amoros and M. Amoros

Color and symmetry
 Arthur L. Loeb

Color and symmetry

Arthur L. Loeb

Ledgemont Laboratory
Kennecott Copper Corporation
Lexington, Massachusetts

and

Department of Visual and Environmental Studies
Harvard University
Cambridge, Massachusetts

Wiley-Interscience, a division of
John Wiley & Sons, Inc.,   New York · London · Sydney · Toronto

Library of Congress Catalog Card Number: 78–136718

ISBN 0 471 54335 7

Printed in the United States of America

10 9 8 7 6 5 4 3 2 1

To the memory of my grandfather
Arthur Isaac
whose early influence pointed the way

Foreword to the series

In the 1920's comparatively few people paid much attention to the branch of science known as crystallography, and the only journal devoted to it was circulated to a few hundred subscribers, mostly libraries. Scientists in the classical disciplines hardly recognized crystallography as a science, although each apparently regarded it as a small segment of his own field. No American university boasted a professor of crystallography and any instruction given was adjunct to mineralogy. Nevertheless, papers of interest to crystallographers appeared in journals of mineralogy, physics and chemistry, although not in large numbers. In those days one might claim to have an all-around acquaintance with crystallography because he was able to keep up with the literature.

This is no longer true. In the period between the first and second world wars, science flourished, and scientists not only published more papers, but an increasing proportion of them dealt with solid materials. It was inevitable that the chemists, metallurgists, physicists and ceramists should make increasing use of crystallographic theory and methods, that the journals of many fields should publish more papers of crystallographic interest, and that new journals devoted to crystallography and the "solid state" should arise. Soon the abstracting journals contained hundreds of titles of crystallographic interest with each issue, and now few of us can keep even reasonably well informed about the many aspects of the science of crystals, to say nothing of keeping abreast of the advances in all these aspects. Not only is it out of the question to keep up with the mass of literature that is turned out, but it is even a little difficult to maintain contact with all the advances in one's own specialty. Accordingly, we are tending to become parochial.

In the words of Warren Weaver"...the volume of the appreciated but not understood keeps getting larger and larger." In order to improve this condition to some extent we need the services of those who, having become authorities in some segments of our field, are willing to integrate their understandings of these limited regions. With such help many of us can gain a sufficient understanding of matters whose original literature we have neither the time nor the inclination to study. Such writings exist in several fields, but none, to date, in crystallography. It is to fill this need that the Wiley Monographs in Crystallography are offered.

MARTIN J. BUERGER

Foreword

When Arthur Loeb first asked me to write a foreword to the book that I had been watching develop, it seemed vaguely like a very nice idea. On further reflection, I began to wonder just what a foreword was supposed to do for a book that the author couldn't do himself. The shelves of my technical library gave little guidance: although a scientific author may have explained the purpose of his work in a preface, he generally worried along without any foreword at all. However, when I turned to more literary and literate volumes, I found the foreword an institution with tradition and purpose. Here, it appeared, an appreciative bystander might orient the reader as to the position of the book, not just within the narrow limits of its subject, but in the world as a whole.

Taking this as my charge, I must begin by calling this work a thing of beauty. The esthetic appeal of symmetrical patterns lies deep and old in our civilization. And we see here a fulfillment of man's eternal search for order, in the characterization—understanding if you will—of the infinite variety of color patterns. But beyond that, Loeb's method and execution is a thing of classical beauty in itself. He moves through the steps of its logical analysis with the same grace and precision as he does through the formal steps of the Renaissance dances which he loves with equal fervor. Here, in the symmetry of both mathematics and the dance, is a beauty of completeness, but one that is so fundamental that it never loses the freshness of its appeal each time we come back to it.

The seven line groups and the seventeen plane groups have been enumerated many times, in many different ways, over the past century. Loeb's method here is a fresh one — mathematical but not conventional. The crystallographer or mathematician familiar with other derivations of the infinite plane groups will find little he can skip over. Loeb's "algorismic" approach defines a bare minimum of symmetry elements—rotocenters—and examines in logical sequence the implications of their coexistence; mirrors, glide lines, and infinite nets are consequences. So are continuous sets of rotocenters, as well as infinite-fold rotations, cases not usually treated. And while the very useful concept of the fundamental (plane-filling) region is developed, the system almost loses sight of the translation lattices as usually defined.

The conventional classification and symbolism of group symmetry are necessarily replaced by others resulting from Loeb's viewpoint.

The reason for this unconventional approach becomes evident in the later chapters, when it serves to simplify the newer concepts of color rotations, and their combinations and their consequences. This of course is the main branch of the work, bursting into lovely flower in the multicolor patterns. I should also like to add that Loeb's logic gives good reason for discarding Shubnikov's idea of "gray" groups — a concept that always left me a little uneasy. The enumeration of plane color groups concludes with a detailed analysis of some of M. C. Escher's drawings. These are fascinating in their symmetry, color, and content, although we might have hoped for an extension of the perceptive esthetic judgment with which Loeb concludes the treatment of mono- chromatic groups.

This is a work that is esthetically and intellectually pleasing. I hope the readers will join me in urging Arthur Loeb to extend his imaginative talents beyond Flatland, and give us a sequel in three (or more) dimensions.

WILLIAM T. HOLSER

Chevron Oil Field Research Company
La Habra, California
March 9, 1970

Preface

The repetition of patterns, whether visual or aural, spatial or temporal, is at the root of existence. Rhythm is the essence of life; we measure the passing of time by the oscillations of pendulums or molecules. Artists and designers have been fascinated for centuries by the repetition of identical modules (Figs. 1, 83, 100, 101, 102). However, the study of crystals, particularly since the invention of x-ray diffraction, has given a particular strong impetus to the mathematical study of repeat patterns.

The discussion in this book is confined to the euclidean plane, but it is concerned with three dimensions: two geometrical coordinates and color. Color is defined more generally than a geometrical coordinate, but in a restricted sense it could be so interpreted. All crystal structures have planar periodic projections and cross sections, which must obey the restrictions discussed in this monograph. In fact, all except the cubic crystal classes are usually considered as stacked two-dimensional arrays. The study of two-dimensional periodic arrays is therefore the basis for geometrical crystallography; color can, among other functions, serve as a "stacking" function.

The electronic computer has brought about a change in philosophy about exhaustive enumerations. Previously, such enumerations were made independently by several workers (e.g., Federov, Schoenflies, and Barlow in the case of the space groups), and when their results agreed it was assumed that the enumeration was indeed complete. Sometimes a mathematical proof would follow, as in the case of the plane symmetry groups, whose completeness was proven topologically by Steiger in 1936. Computers can make very extensive and dependable enumerations once they are given accurate well-defined instructions; the rule for generating an enumeration (or, in general, a function) is called an algorism. Without employing a computer, we have used such an algorismatic approach in our enumeration of all planar colored configurations. Here it is not so much a question of *proving* exhaustiveness as of clearly defining the algorisms and then generating the configurations exhaustively according to these rules.

The method used is thus synthetic rather than analytical. Configurations are not at once presented in toto but are grown organically. The coexistence of two symmetry elements in the plane is found to be subject to constraints; if these constraints are satisfied, the coexisting elements usually imply a third element, which in turn interacts with the original two to imply further elements, continuing until a complete configuration is generated. Each configuration is thus characterized by the original pair (sometimes trio) of symmetry elements from which it was generated. It is convenient therefore to classify these configurations according to their generating elements; a nomenclature corresponding to this classification will be presented.

The investigation reported on in this monograph grew out of a reexamination of the so-called seventeen plane symmetry groups, inspired by and done in collaboration with Professor Philippe LeCorbeiller. Geographical separation made a continued collaboration impractical, but if any virtues be found in this book, the responsibility for them is likely to be Professor LeCorbeiller's. I myself accept the responsibility for the many shortcomings.

The invitation from Professor Martin Buerger to write this monograph was not always appreciated during the writing of it, but his encouragement and critical assistance have made the task a light one. The exemplary good cheer with which Mrs. Evalyn Sorrentino typed the manuscript has been an inspiration.

I am indebted also to Professor C. Wroe Wolfe for a critical examination of the original report on this investigation.

The publications group at Ledgemont Laboratory, headed by Roland Johnson, has been most helpful in coping with many technical problems in the production of the monograph. Donald Barclay and Jean Hasch have personally seen the colored illustrations through every step from the author's sketches to the final printed page.

The generous support from the Kennecott Copper Corporation in the research and production of the monograph is gratefully acknowledged. I also wish to add a personal note of appreciation for the moral support and friendship experienced from Dr. Ewan W. Fletcher, Director of Ledgemont Laboratory.

ARTHUR L. LOEB

Lexington, Massachusetts
July 1970

Contents

Color and symmetry

1

Patterns, transformations, and symmetry

A. Patterns

A plane pattern is a collection of points in a plane. One pattern is distinguished from another by the relation its points bear to each other. The term "pattern recognition" is frequently used in communication science to mean the recognition, by a human being or by a machine, of a significant relation between the points of a pattern. Crystallographers are concerned with significant geometrical relations between the constituent atoms or ions in a crystal.

B. Transformations

Two patterns can be *transformed* into each other if an operation exists that turns each point P_0 in one pattern into a corresponding point P_1 in the other one, and if a second operation exists that turns each point P_1 into the corresponding point P_0. These two operations, when applied successively in either sequence, leave the pattern unchanged; such operations are said to be *inverses* of each other.

As an example of a transformation we shall consider *scaling*. A pattern is transformed into a second one by scaling if any three non-linear points P_0, Q_0, and R_0 in the original pattern correspond to three points P_1, Q_1, and R_1 in the second pattern such that

$$\frac{\overline{P_0Q_0}}{\overline{P_1Q_1}} = \frac{\overline{Q_0R_0}}{\overline{Q_1R_1}} \quad \text{and} \quad \angle\, P_0Q_0R_0 = \angle\, P_1Q_1R_1.$$

1

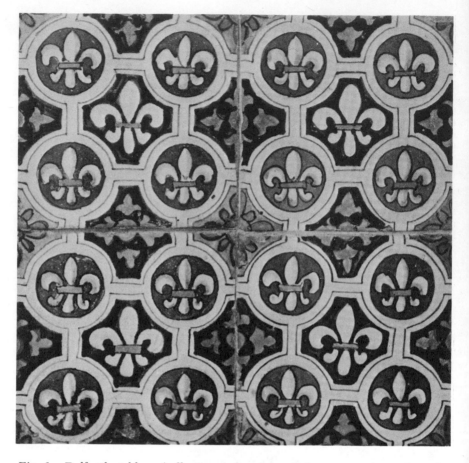

Fig. 1. Delft tile tableau (collection Arthur Isaac, Ryksmuseum, Amsterdam, the Netherlands).

These equations express the fact that in scaling all linear dimensions are changed by the same factor, and all angles remain the same. When this factor equals unity, so that all line segments retain the same length, the transformation is called a *coincidence* transformation. The triangles $\triangle P_0 Q_0 R_0$ and $\triangle P_1 Q_1 R_1$ in Fig. 2 are *similar* when they are related by a *scaling* operation; they are *homometric* when related by a *coincidence* operation. The homometric triangles are either *congruent* or *enantiomorphs*; in the former case the coincidence operation is *direct*, in the latter it is *opposite*.

A *direct coincidence* operation that leaves a single point unaffected is called a *rotation*. The invariant point is called a *rotocenter*. A *direct*

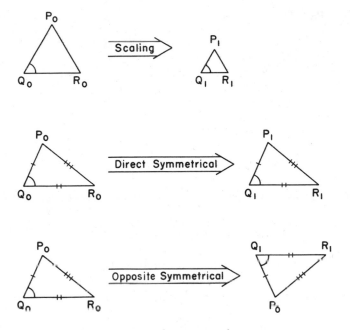

Fig. 2.   Scaling and symmetrical operations.

coincidence operation that turns each segment P_0Q_0 into an equally long segment P_1Q_1 parallel to P_0Q_0 is called a *translation* (Fig. 3).

Any two scalene congruent triangles may be transformed into each other by a single rotation. The rotation becomes a translation in

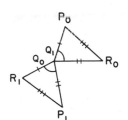

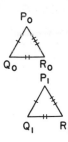

Rotation : $Q_0 = Q_1$ = Rotocenter

Translation : $\overline{P_0Q_0} \parallel \overline{P_1Q_1}$
$\overline{Q_0R_0} \parallel \overline{Q_1R_1}$
$\overline{R_0P_0} \parallel \overline{R_1P_1}$

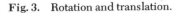

Fig. 3.   Rotation and translation.

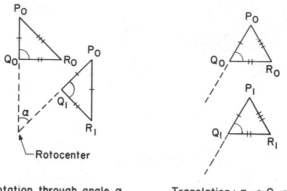

Rotation through angle α Translation: $\alpha \rightarrow 0$, rotocenter
                                                  at infinity

Fig. 4.   Rotation as the general symmetrical direct operation.

the limiting case when the rotocenter lies at infinity (Fig. 4). In this
treatment of coincidence we shall consider rotation as the most general
direct coincidence operation and translation as a special case of rotation.

A coincidence operation that turns every scalene triangle $P_0 Q_0 R_0$
into an enantiomorphic triangle $P_1 Q_1 R_1$ is called a *glide reflection*. The
midpoints of the line segments $\overline{P_0 P_1}$, $\overline{Q_0 Q_1}$, and $\overline{R_0 R_1}$ lie on a straight
line called the *glide line* (Fig. 5). Every point S_0 located on the glide
line is transformed into another point S_1 on the same line by the glide
transformation. The length of the line segment $S_0 S_1$ is called the *translation component* of the glide operation.

In the special case when the *translation component equals zero*,
the reflection operation is called *mirror reflection*. The glide line
becomes a *mirror line* in this special instance; mirror reflection leaves
every point on the mirror line unaffected (Fig. 6).

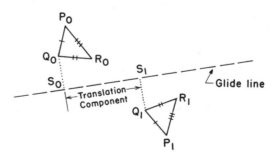

Fig. 5.   Glide reflection.

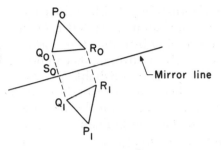

Fig. 6.   Mirror reflection.

We have thus defined two basic coincidence operations: a direct operation called *rotation* (a single point left unchanged; the special case of *translation* results when this point goes to infinity), and an opposite operation called *glide reflection*, or simply *glide* (a single line is unchanged; the special case of *mirror reflection*, or simply reflection, results when every point on that line is unchanged).

The trivial operation that leaves every point in the plane unaffected is called the *identity operation*. The application in succession of two operations that are each other's inverses is equivalent to the identity operation. The identity operation plays a role in symmetry theory similar to that of multiplication by unity or the addition of zero in algebra.

C.   Symmetry

The points in a pattern may be intrinsically different from each other, or they may be distinguishable from each other only by the geometrical relationships that they bear to each other. We first consider patterns made up of intrinsically identical points, noting geometrical relationships only. Later we shall assign colors to the points to symbolize the intrinsic properties in which they differ from each other.

Two points that are intrinsically identical and bear identical geometrical relationships to all points in their pattern are indistinguishable from each other. If a pattern contains two such points, P_0 and P_1, then a coincidence operation that transforms P_0 into P_1 results in a pattern indistinguishable from the original pattern. When a coincidence operation transforms a pattern into another pattern that is indistinguishable from the original one, this pattern is said to be *symmetrical*, or to possess *symmetry*. Every symmetrical pattern necessarily contains at least two points that are related by a symmetrical transformation and its inverse.

A complete set of indistinguishable points, each of which is related to every other by translation, is called a *lattice*. Indistinguishable points related to each other by either translation or rotation are called *congruent*, because their environments can be made to coincide by a direct operation. Two indistinguishable points related to each other by glide or mirror reflection are called *enantiomorphs*. Points that are *either* congruent *or* enantiomorphs are called *equivalent*. Points that are not equivalent, i.e., neither congruent nor enantiomorphs, are called *distinct*. Schematically, points can be distinguished as follows:

congruent

enantiomorphic } equivalent

noncongruent

distinct

2

Rotational symmetry

A. Patterns with not more than a single rotocenter

Rotation, as we noted previously, leaves a single point, the roto-center, unaffected. In a pattern that has rotational symmetry every point not coincident with a rotocenter belongs to a set of congruent points distributed at equal angular intervals over the circumference of a circle centered on the rotocenter. The number of congruent points on this circumference will be called the *symmetry number* of the rotocenter (Fig. 7*a*).

In Chapter 1 we observed that a pattern is characterized by the geo-metrical relationship its points bear to each other. In Fig. 7*b* we see

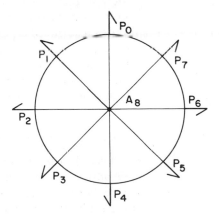

Fig. 7*a*.   Rotocenter A_8 is a center of eightfold symmetry. The points P_0 to P_7 are all congruent.

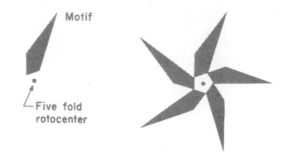

Motif

Five fold
rotocenter

Fig. 7*b*.   Generation of a pattern from an asymmetrical motif by a fivefold rotocenter.

how a pattern is *generated* by the symmetrical relationship imposed on an asymmetrical motif by a fivefold rotocenter. We say that the rotocenter *acts on* the motif to *generate* the pattern, in this case a five-pointed star.

Let us divide a plane that contains a k-fold rotocenter (i.e., a rotocenter of symmetry number k) into k wedge-shaped regions joining at the rotocenter. Any point in any such region is congruent to an equivalent point in each other region.

Two special cases are considered separately. First, when $k = 1$, no two points in the plane are congruent to each other. This case represents a pattern without rotational symmetry. Second, when $k = \infty$ and the rotocenter is at infinity, the wedge-shaped regions become parallel zones, and the set of points located on the circumference becomes a series of equally spaced points on a straight line. The pattern of Fig. 8 illustrates this case. *All* equivalent points are here related by translation, so that they constitute a lattice.

B.   Coexistence of rotocenters

Fig. 8.   Translational symmetry: $k = \infty$.

Having defined our vocabulary, we are now ready to venture on our first kaleidoscopic enterprise. "Kaleidoscope" is used here in its most general sense, namely, as revealing patterns not only by means of two or more mirrors but

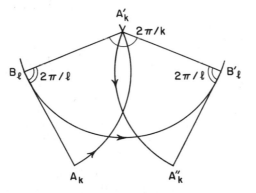

Fig. 9.   Interaction of rotocenters.

by combining rotocenters and reflection lines. In the first instance we place two rotocenters a finite distance apart in the plane. One of these is A_k, a k-fold rotocenter, the other B_l, an l-fold rotocenter. The values of k and l may or may not be the same (Fig. 9).

Because B_l is an l-fold rotocenter, A_k is one of a set of l congruent rotocenters lying on a circle centered on B_l. Of these l rotocenters only two are shown in Fig. 9, the original A_k and A_k'. The latter is chosen such that $A_k B_l A_k' = 2\pi/l$; this means that on the smaller arc $A_k A_k'$ of the circle around B_l there is no other rotocenter congruent to A_k and A_k'.

Similarly, A_k' is surrounded by k rotocenters equivalent to B_l (including B_l itself). One of these is B_l', chosen on the circle of B-s such that there is no other rotocenter equivalent to B_l on the smaller circular arc $B_l B_l'$. Furthermore, the sense of rotation from B_l to B_l' around A_k' is the same as that from A_k to A_k' around B_l.

In turn, we can center our attention on B_l', which is surrounded by l rotocenters, of which two are shown, A_k' and A_k''. The sense of the rotation around B_l' is again the same as that used in the previous operations, and no other rotocenter equivalent to A_k' is to lie on the smaller circular arc $A_k' A_k''$.

Will this process continue indefinitely, or will one of the A_ks, say $A_k^{(i)}$, eventually coincide with A_k? We shall see presently that this depends on the values of k and l. Before attempting a general solution to this problem, however, we should examine the patterns that might possibly be generated. We begin by examining some representative sample values of k and l, and subsequently we return to the general problem.

The first example is $k = 3$, $l = 6$, illustrated in Fig. 10a. Here B_6

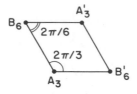

Fig. 10*a.* Interaction of a threefold and a sixfold rotocenter.

Fig. 10*b.* Interaction of a motif having threefold rotational symmetry with a sixfold rotocenter.

acts on A_3 to generate a rotocenter coincident with A_3'; a closed polygon (in this case an equilateral parallelogram) is generated. In Fig. 10*b* we have centered a motif with threefold rotational symmetry on a threefold rotocenter; a sixfold rotocenter nearby generates a second motif, equivalent to the first. Later we shall extend this pattern further; for the moment we limit ourselves to an example of the general Fig. 9, in which only one rotocenter was generated at a time. Figures 10*a* and 10*b* illustrate a very important principle in symmetry theory: the two equivalent triangles are related by a twofold rotocenter that has appeared between them! The reader should now generate some patterns himself, by centering a motif with sixfold rotational symmetry on the appropriate rotocenter in the presence of a threefold rotocenter or by having an asymmetrical motif acted on successively by a threefold and a sixfold rotocenter. In each case a twofold rotocenter will automatically appear and should be located.

Fig. 11*a.* The interaction of two fourfold rotocenters.

The second example to be considered has $k = l = 4$, and is illustrated in Figs. 11*a* and 11*b*. As shown in Fig. 11*a*, a closed polygon (in this case a square) is generated once more. The pattern used to illustrate this case (Fig. 11*b*) shows clearly that the two original rotocenters, though having the same symmetry value, are quite distinct. A twofold rotocenter may be observed at the center of the square $A_4B_4A_4'B_4'$.

The third example, illustrated in Figs. 12*a* and 12*b*, has $k = l = 2$. Here a straight line is generated, along which the two distinct rotocenters alternate at equal intervals. For these values of k and l the rotocenters do not constitute the vertices of a closed polygon; instead, the sequence $A_2B_2A_2'B_2'A_2 \ldots$, lies on a straight line, which extends to infinity. Figure 12*b* was generated by placing an asymmetrical triangular motif with a corner on one twofold rotocenter, having it acted on first by that particular roto-

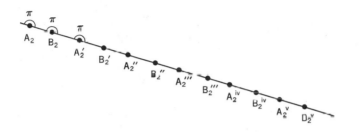

Fig. 11b. The interaction of a motif having fourfold rotational symmetry with a four-fold rotocenter.

Fig. 12a. The interaction of two twofold rotocenters.

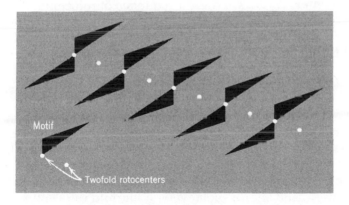

Fig. 12b. The interaction of an asymmetrical motif with two distinct twofold roto-centers.

center, A_2, and then by the second rotocenter, B_2. (The order in which these interactions occur is immaterial.) The resulting pattern also has, besides the original twofold rotational symmetry, *translational* symmetry; we noted before that translational symmetry may be considered the limit of k-fold rotational symmetry as $k \to \infty$.

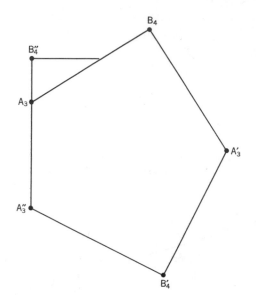

Fig. 13a. The interaction of a threefold and a fourfold rotocenter (initial stages).

Fourth we consider $k = 3$, $l = 4$, illustrated in Fig. 13a. Starting with rotocenters A_3, B_4, we trace a sequence of rotocenters generated by each other: $A_3B_4A_3'B_4'A_3''B_4'' \ldots$. Before continuing, let us stop to examine this point B_4'' and observe that it lies much closer to A_3 than the original B_4 was. Let us call the ratio of these distances

$$\frac{\overline{A_3B_4''}}{\overline{A_3B_4}} \equiv r.$$

Instead of continuing the polygonal line beyond B_4'', as shown in Fig. 13b, we could have started all over again with the pair of rotocenters A_3, B_4'', and by repeating the procedure analogously, we would create a rotocenter congruent to B_4 at a distance r^2 times the original distance $\overline{A_3B_4}$. Each time the newly created fourfold rotocenter could be combined with A_3, and after n such cycles a fourfold rotocenter would be generated at a distance $r^n\overline{A_3B_4}$. Since there is no limit to the number of times that this procedure can be repeated, and since $r < 1$, a fourfold rotocenter will eventually be generated that is arbitrarily close to A_3. Thus the coexistence in a plane of a threefold and a fourfold rotocenter implies the existence of rotocenters arbitrarily close to each other in that plane.

In principle, this continuous distribution of rotocenters through the plane, though unfamiliar, is not forbidden. It is not a situation that

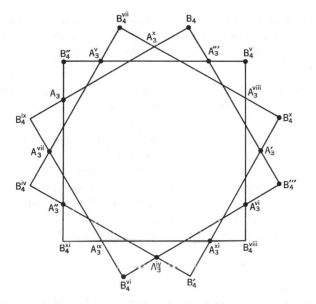

Fig. 13b. The interaction of a threefold and a fourfold rotocenter (continued). Note that after five cycles the polygonal line closes upon itself: $A_3{}^{x12} \equiv A_3$.

occurs in crystallography or in mosaic patterns, though. We shall arbitrarily exclude continuous distributions of rotocenter through a postulate of closest approach; this postulate limits us to consideration of *discrete* patterns. The study of *continuous* patterns, while also valid and interesting, is outside the scope of this book.

C.   A fundamental postulate

This postulate is basic to our discrete symmetry theory; it is needed for proving a great many theorems. It must be realized, therefore, that such theorems are not necessarily true for patterns having continuous distributions of rotocenters.

It will be convenient for a statement of this postulate to expand our vocabulary. We have noted that the coexistence of a pair of distinct rotocenters A_k, B_l implies a set of rotocenters A_k, A_k', A_k'', ..., and a set B_l, B_l', B_l'', ...; in the examples considered, additional symmetry was observed. We can expect, therefore, to find large numbers of rotocenters, some of which (e.g., A_k and A_k') are congruent to each other, while others (e.g., A_k and B_l) are distinct. All rotocenters congruent to each other form a set which will be called a *rotosimplex*. In these terms we can state this fundamental postulate.

Postulate of closest approach. For every pair of rotosimplexes in a plane there exists a finite distance such that no point in one simplex is closer to any point of the other simplex than that given distance.

The example $k = 3$, $l = 4$ was shown to violate this postulate, hence it is excluded. The postulate of closest approach makes no mention of distances *within* a rotosimplex. Such a statement need not be made explicitly, for if rotocenters belonging to the same rotosimplex were allowed to approach each other arbitrarily closely, then they would be continuously and uniformly distributed over the plane. Accordingly, some such rotocenters would be located arbitrarily closely to some rotocenters belonging to a second rotosimplex and hence violate the postulate of closest approach. From this postulate it follows therefore that no pair of equivalent rotocenters may be arbitrarily close to each other.

We can now generalize the principles illustrated by the examples of the previous section. Let us reconsider Fig. 9 as redrawn in Fig. 14 in the light of the new postulate (we drop the subscripts for convenience). We can assume that A and B were chosen such that no rotocenter equivalent to B is closer to A than B. Bisect the angles $\angle ABA'$ and $\angle BA'B'$; unless $k = l = 2$, the bisectors intersect at a finite point C. Join this point C to B'. Since $\overline{BA'} = \overline{B'A'}$, $\angle BA'C = B'A'C$, and $\overline{CA'} = \overline{CA'}$, $\triangle CBA'$ is oppositely congruent to $\triangle CB'A'$. Therefore $\overline{CB} = \overline{CB'}$,

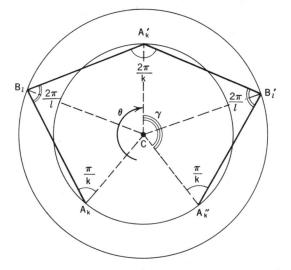

Fig. 14.   Generation of a polygonal line by the interaction of two rotocenters.

and $\angle A'B'C = \angle A'BC = \pi/l$. Since B and B' are related by rotational symmetry, $\angle A'B'A'' = \angle ABA' = 2\pi/l$. Thus it is proven that the line CB bisects $\angle A'B'A''$ and that C is equidistant from B and B'. It is proven similarly that C is equidistant from A, A', A'', etc. The sequence line $ABA'B'A''\ldots$, therefore alternates between two circles, both centered on C. The following are the courses that this line sequence could run:

1. It could extend to infinity. This would happen if $k = l = 2$; in this case the bisectors would all be parallel to each other, and perpendicular to the straight line $A_2B_2A_2'B_2'\ldots$, so that the point C would lie at infinity (e.g., Figs. 12a and 12b).

2. The sequence would close upon itself after more than one, but a finite number of revolutions around C (e.g., Fig. 13b).

3. The sequence would continue indefinitely to circle around C, never closing on itself.

4. The sequence would close upon itself after a single revolution (e.g., Figs. 10 and 11).

In considering these possibilities, let $\angle ACA' = \angle BCB' = \angle A'CA'' = \ldots \equiv \gamma$. The relative angular positions about C of A, A', $\ldots$, $A^{(i)}$, $\ldots$, are, respectively, 0, γ, $\ldots$, $i\gamma$, $\ldots$; while those of B, B', $\ldots$, $B^{(i)}$, $\ldots$, are $\frac{1}{2}\gamma$, $\frac{3}{2}\gamma$, $\ldots$, $(i+\frac{1}{2})\gamma$, $\ldots$.

Let the angular position about C be generally denoted by θ. Then if there exists a rotocenter $A^{(i)}$ in the interval $2\pi - \gamma < \theta < 2\pi$, there is a rotocenter $B^{(i)}$ in the interval $2\pi - \frac{1}{2}\gamma < \theta < \frac{3}{2}\gamma$; this rotocenter is closer to A than B is, and therefore violates the postulate of closest approach. If there is a rotocenter $A^{(i)}$ equivalent to A in the interval $(2\pi - 2\gamma) < \theta < (2\pi - \gamma)$ then there must be a rotocenter $A^{(i+1)}$ in the interval $(2\pi - \gamma) < \theta < 2\pi$, but this was just shown to lead to a violation of the postulate. Therefore there can only be a rotocenter $A^{(i)}$ at $\theta = 2\pi - \gamma$; then $A^{(i+1)} = A$, so that the polygonal line closes upon itself after a single rotation. The cases 2 and 3 inevitably produce rotocenters in the forbidden intervals, and are therefore excluded by the postulate of closest approach, as was exemplified explicitly by the example $k = 3$, $l = 4$. In the next chapter we shall find explicitly which combinations of k and l are permitted by the postulate.

3

The first diophantine equation

A.   Implication of symmetry

Since the sequence $ABA'B'A''\ldots$, was shown to constitute the vertices of a polygon, and since $A, A', A''\ldots$, as well as $B, B', B'',\ldots$, are related by rotational symmetry, the point C is itself a rotocenter. It corresponds to the twofold rotocenters that appeared automatically in the examples $k=3$, $l=6$, and $k=l=4$. Since the existence of C as a rotocenter is a direct result of the coexistence of A and B in the same plane, we say that A and B interact to imply C. The symmetry value of C is found from the quadrilateral $A'_k B'_l A''_k C$:

$$\frac{\pi}{k}+\frac{2\pi}{l}+\frac{\pi}{k}+\gamma = 2\pi$$

$$\therefore \gamma = 2\pi\left(1-\frac{1}{k}-\frac{1}{l}\right).$$

If we use m to designate the symmetry value of the rotocenter C, then $\gamma = 2\pi/m$, so that

$$\frac{1}{k}+\frac{1}{l}+\frac{1}{m}= 1. \tag{1}$$

This equation expresses Theorem 1.

Theorem 1. The coexistence in a plane of two rotocenters implies the existence of a third rotocenter in the same plane. The symmetry values k, l, m of the triplet of rotocenters so related are constrained by the equation $(1/k) + (1/l) + (1/m) = 1$.

16

Equation (1) is called a diophantine equation after the Alexandrian mathematician Diophantos. In a diophantine equation all variables are limited to rational values; in our case the variables are integers. Our diophantine equation is unaffected by any permutation of the variables k, l, and m: this equation is *symmetrical* in k, l, and m. As a result, we can permute the subscripts k, l, and m in the proven statement "A_k and B_l interact to imply C_m" to read

$$A_l \text{ and } B_m \text{ interact to imply } C_k$$
$$A_m \text{ and } B_k \text{ interact to imply } C_l$$
$$A_m \text{ and } B_l \text{ interact to imply } C_k \text{ etc.}$$

These statements are illustrated in Fig. 15. Here Fig. 10b is extended by surrounding each sixfold rotocenter by *six* threefold motifs and by surrounding each threefold motif by *three* sixfold rotocenters. The result is an infinitely extended pattern, of which Fig. 15 shows a finite portion. Several different polygons could have been used as starting point for Fig. 15: the interaction of A_3 and B_6 that was illustrated in Figs. 10 as well as the interaction between A_2 and B_3 (C_6 implied) and between A_6 and B_2 (C_3 implied).

It should be noted that combinations $k = 3$, $l = 6$, $m = 2$ (Figs. 10 and 15) and $k = 4$, $l = 4$, $m = 2$ (Fig. 11) obey the diophantine equation. Substitution of $k = 3$, $l = 4$ into (1) leads to $m = 12/5$, a result that could be interpreted in terms of Fig. 13b to mean that $A_3^{\text{xii}} \equiv A_3$ after five cycles around C. However, such a fractional value of m violates the postulate of closest approach: k, l, and m are limited to integer values. We shall now find systematically which combinations of k, l, and m are permitted.

B.   Solution of the first diophantine equation

We have noted that (1) is symmetrical in k, l, and m: of the triplet A_k, B_l, C_m any two imply the third one. Therefore any permutation of numerical values between k, l, and m does not lead to an intrinsically distinct solution. Thus without loss of generality we can assume that $k \leqslant l \leqslant m$.

1. Solutions having $k = 1$. When $k = 1$ is substituted in (1),

$$\frac{1}{l} + \frac{1}{m} = 0,$$

$$\therefore l = \infty \quad \text{and} \quad m = \infty.$$

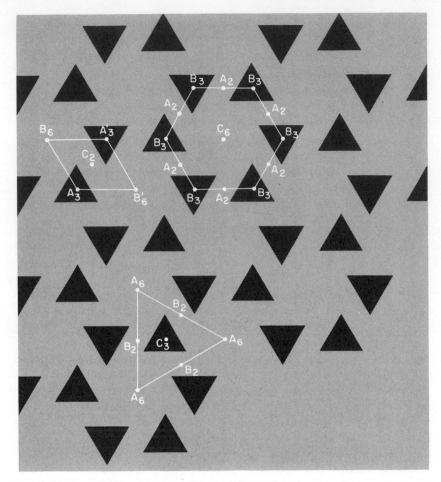

Fig. 15. This pattern can be considered equally as the result of the interaction between an A_2 and a B_3, an A_3 and a B_6, or an A_6 and a B_2.

2. Solutions having $k = 2$. When $k = 2$ is substituted in (1),

$$\frac{1}{l} + \frac{1}{m} = \frac{1}{2}.$$

Since $l \geqslant k$, we have the following possibilities:

$$l = 2, \qquad m = \infty$$
$$l = 3, \qquad m = 6$$
$$l = 4, \qquad m = 4$$

If $l > 4$, $m < 4$, $\therefore m < l$. Since this violates the assumption $l \leq m$, $l \leq 4$.

3. Solutions having $k = 3$. When $k = 3$ is substituted in (1),

$$\frac{1}{l} + \frac{1}{m} = \frac{2}{3}.$$

Here $l \geq 3$, hence the possibilities are:

$$l = 3, \quad m = 3$$
$$l > 3, \quad m < 3$$

The second possibility violates the condition $l \leq m$, so that when $k = 3$, l and m are limited to the value 3.

4. Solutions having $k > 3$. In this case

$$\frac{1}{l} + \frac{1}{m} = 1 - \frac{1}{k} > \frac{2}{3}.$$

Since $l \geq k$ and $m \geq k$, $l > 3$ and $m > 3$ in this case. This would make it impossible to make $(1/l) + (1/m) > \frac{2}{3}$. Therefore $k \leq 3$.

There are therefore five solutions to the first diophantine equation, summarized in Table 1.

Table 1   Solutions of the
first diophantine equation

k	l	m
1	∞	3
2	2	∞
2	3	6
2	4	4
3	3	3

C.   Nets

In the construction of Fig. 15 it was noted that the interaction of two rotocenters eventually generates an infinite pattern of rotocenters. These rotocenters belong to three distinct rotosimplexes; two of these contain one of the original rotocenters each.

If lines are drawn so that every rotocenter is joined to its nearest neighbors in each of the other two simplexes, a net is formed whose nodes are rotocenters. The nets for each of the solutions listed in Table 1 will now be considered, starting at the bottom and working upward to the solutions having infinite symmetry values.

Figure 16 corresponds to $k = l = m = 3$. It was generated by placing larger triangles around a smaller one with the requirement that the center of each triangle be a threefold rotocenter for the entire pattern. The implied third rotosimplex is located at the centers of irregular hexagons formed by the interspace between the triangles. These irregular hexagons all have threefold rotational symmetry. Figure 16 also shows the net formed by the three distinct threefold roto-simplexes.

Figure 17 illustrates $k = 2$, $l = m = 4$. This pattern resulted from a motif with twofold rotational symmetry and a fourfold rotocenter at a corner of this motif. The net of rotocenters is superimposed on this pattern, showing that the second fourfold rotosimplex emerges between the twofold motifs. Figure 11, which also illustrates this solution, can be extended in order to find the symmetry value and location of the implied rotocenters.

Figure 18 illustrates the solution 236, as did Fig. 15. Since we want to show the net for the solution 236, a second illustration was designed for this example. This time a twofold motif was used in combination with a threefold rotocenter, with the result that a sixfold rotosimplex emerges in the spaces between the threefold motifs. The two ex-amples of Figs. 15 and 18 and also of Figs. 11 and 17 demonstrate

Fig. 16. Pattern and net representing the solution $k = l = m = 3$. The three roto-simplexes are labeled 3, 3' and 3" respectively.

Fig. 17.   Pattern and net representing the solution $k = 2, l = m = 4$.

that there is an infinite variety of symmetrical *patterns* that can be designed in the plane. We shall find, however, that these patterns correspond to a strictly limited number of *nets*, i.e., configurations of symmetry elements.

For the solution 22∞ recall Section 2.C, where we examined the various possible courses that the polygonal line $ABA' \ldots$ could take. Of the four possibilities, two were rejected on the basis of the postulate of closest approach, and two were retained. Of the two that were retained, one led to a closed polygon and the diophantine equation while the other was one where $k = l = 2$; the line sequence here became a straight line that did not close upon itself. It was noted that the coexistence of two twofold rotocenters implied translational, i.e., infinite-fold rotational symmetry. We see now that this description perfectly fits the solution 22∞ of the diophantine equation: this equation therefore covers *both* possible courses of the line sequence. The "net" corresponding to the solution 22∞ is simply the straight line $A_2 B_2 A_2' B_2' \ldots$, with regularly spaced nodes at the twofold rotocenters.

Fig. 18. Pattern and net representing the solution $k = 2$, $l = 3$, $m = 6$ (cf. Fig. 16).

From these nodes straight parallel lines emanate to join these twofold rotocenters to the implied C_∞ at infinity (Fig. 19). This solution is illustrated by the pattern drawn in Fig. 12b.

The remaining solution of the diophantine equation, 1∞∞, has no

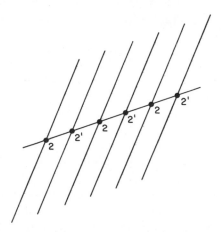

Fig. 19.   Net for the solution 22∞.

finite rotational symmetry: the symmetry value $k = 1$ implies the total absence of rotational symmetry. Since infinite-fold rotational symmetry amounts to translation symmetry, the two ∞s in the solution indicate translational symmetry in two independent, i.e., nonparallel, directions. This symmetry is illustrated in Fig. 20. Since there are no finite rotocenters, there are no nodes and no net.

D.   Fundamental regions and meshes

From the postulate of closest approach it follows that every point of a plane has a finite region associated with it within which no other point is equivalent to it. Two regions are said to be equivalent to each other when every point in one region is equivalent to a point in the other region. The plane can be entirely covered by mutually equivalent regions; when *no* two points within one such region are equivalent, this region is called a *fundamental region*. Generally, a fundamental region is not uniquely defined, although its area always is: in the absence of enantiomorphy its outline can be arbitrarily determined. In a pattern that has translational symmetry only, a fundamental region is identical with a unit cell. When a pattern has rotational symmetry as well, a unit cell contains several equivalent rotocenters; the unit cell can then be equally divided between these rotocenters, giving each an equivalent portion, which constitutes a fundamental region for the pattern.

Fig. 20.   Pattern illustrating the solution $1\infty\infty$, in which there is translational symmetry in two nonparallel directions.

The nets whose nodes constitute the rotocenters of a given pattern divide the plane into regions which are, in the language of network theory, called *meshes*. These meshes are *not* synonymous with two-dimensional unit cells, which are sometimes given the same name.

The number of meshes meeting at any rotocenter equals *twice* the symmetry value of that rotocenter. The number of congruent points at a given distance from a rotocenter equals the symmetry value of that rotocenter. Therefore two points in adjacent meshes (meshes that share an edge) cannot be congruent, though they may be enantiomorphs.

Any pair of meshes in a net is either equivalent or distinct. Two adjacent meshes are *either* enantiomorphs or *distinct*, but *never* congruent. (Note, for instance, Fig. 16: although the *outlines* of adjacent meshes are geometrically congruent, their contents are not.) When a pair of adjacent meshes are distinct, they constitute together a fundamental region. In Fig. 18, a fundamental region may be formed in three different ways from pairs of adjacent meshes: the three resulting regions all have different shapes, but the same area.

E.   Summary

The following are the combinations of symmetry values found so far:

A single k-fold rotocenter — special cases:

$k = 1$:   no rotational symmetry
$k = \infty$:   translational symmetry in a single direction

Three rotosimplexes:

$k = 1$, $l = m = \infty$: translational symmetry in two nonparallel directions
$k = l = 2$, $m = \infty$: a linear array of alternating two-fold rotocenters
$k = 2$, $l = 3$, $m = 6$ ⎫
$k = 2$, $l = m = 4$ ⎬ two-dimensional nets of rotocenters
$k = l = m = 3$ ⎭

4

An upper limit on the number of permitted rotosimplexes

A. Finite symmetry values

Having considered patterns with a *single* rotocenter (any integral symmetry value permitted), and the interaction of two generally non-congruent rotocenters (symmetry values limited by the diophantine equation), we now proceed to evaluate the interaction of *any three* rotocenters. According to Theorem 1, the coexistence of two roto-centers implies a third rotocenter; a net of infinite extent is eventually generated by the continued interactions of the rotocenters generated. If the third postulated distinct rotocenter coincides with one of the rotocenters implied by the other two, then one of the nets generated in Chapter 3 results, and the third distinct rotocenter is tautological, though, of course, necessary for completeness. We wish to investigate here the extent to which one may postulate three *independent* roto-centers in the plane.

Let us consider first the nets with finite symmetry, 236, 244, and 333. A mesh of such a net is drawn diagrammatically in Fig. 21; A_k, B_l, and C_m are the three rotocenters coexisting by virtue of Theorem 1, and D_p is an additional center which may be congruent with neither A_k, nor B_l, nor C_m. The nodes of the net generated by A_k, B_l consist of *all* rotocenters implied by and implying A_k and B_l. No rotocenters equivalent to or implied by A_k and B_l exist in the plane that are not nodes of this net. The fact that the number of rotocenters in the plane, though infinite, is enumerable is due to the postulate of closest approach.

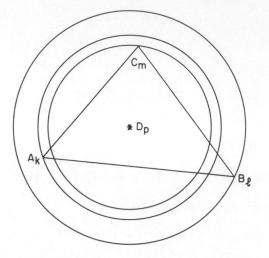

Fig. 21. Rotocenter D_p inside a mesh of the net generated by A_k and B_l.

The p-fold rotocenter D_p implies p rotocenters congruent to A_k (including A_k itself) on a circle centered on D_p having radius D_pA_k and similarly p rotocenters congruent to B_l and p rotocenters congruent to C_m on appropriate circles. If these rotocenters do *not* belong to the net generated by A_k, B_l, the existence of D_p violates the postulate of closest approach. If the rotocenters on the circles centered on D_p do all belong to the net generated by A_k, B_l, then D_p belongs to that net as well, hence it is tautological.

The conclusion is that the triplet A_k, B_l, C_m referred to in Theorem 1 and the diophantine equation is exclusive when k, l, and m are finite: no rotocenters distinct from these can exist in the plane.

B.   The solution 22∞ and D_p

The solution 22∞ of the diophantine equation corresponds to a linear array of alternating, equally spaced twofold rotocenters. This array contains *all* rotocenters collinear and congruent with the originally postulated A_2 and B_2. Note that there is no reason to exclude here rotocenters congruent to A_2 and B_2 but not collinear with these two rotocenters, because such rotocenters would not violate the postulate of closest approach.

If an additional rotocenter D_p is postulated collinear with A_2 and B_2, it may take one of the following three positions.

1. On A_2 or B_2. In this case, D_p would be tautological.

2. Halfway between A_2 and B_2 (or their congruents). If p were even, A_2 and B_2 would be congruent; the postulated distinctness of A_2 and B_2 would then be violated. For odd values of p, see below.

3. Collinear with, but neither *on* nor *halfway* between A_2 and B_2 or rotocenters congruent with those. If p is even, additional rotocenters congruent and collinear with A_2 and B_2 would be generated, so that the postulate of closest approach would be violated.

Since each of these possibilities for D_p is eliminated, there remain the following:

4. D_p is not collinear with A_2 and B_2.
5. D_p is collinear with A_2 and B_2, but p is odd.

The coexistence of A_2 and D_p is limited by Theorem 1 and (1) to the following values of p: $p = 2, 3, 4, 6$. When $p = 3, 4$, or 6, a two-dimensional net will be generated. In Section 4.A we saw that such nets would preclude the existence of rotocenter B_2. Therefore we have systematically eliminated all possible D_p except one: D_p is *not* collinear with A_2 and B_2, and $p = 2$.

C   Four twofold rotosimplexes

The conclusion just reached may be put in the following form:

Theorem 2. The coexistence in a plane of two noncongruent twofold rotocenters precludes the existence in that plane of any rotocenters whose symmetry value does not equal *two*.

Figure 22 contains a linear array $A_2B_2A_2'B_2'$ ..., as well as an additional twofold rotocenter D_2, neither collinear nor congruent with A_2 and B_2. Interaction of D_2 with A_2 produces A_2'', since A_2 and A_2'' are congruent, they must have the same environment, with the result that a linear array $A_2''B_2''A_2'''B_2'''$ ... is implied. The translation symmetry between A_2 and A_2' implies the existence also of D_2', related to D_2 by translation symmetry. Thus an infinite net of *three* interpenetrating twofold rotosimplexes is generated. The lines B_2B_2'' and A_2A_2''' intersect at point E. Since D_2' is a twofold rotocenter, $\triangle B_2A_2'D_2'$ is congruent with $\triangle B_2'''A_2'''D_2'$. Because of

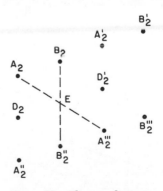

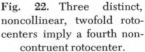

Fig. 22. Three distinct, noncollinear, twofold rotocenters imply a fourth noncontruent rotocenter.

the translational symmetry between D_2 and D_2', $\triangle B_2'' A_2'' D_2$ is congruent also with $\triangle B_2''' A_2''' D_2'$. Therefore $\triangle B_2 A_2' D_2'$ is congruent with $\triangle B_2' A_2'' D_2$; these triangles are transformed into each other by a 180° rotation around E. Any triangle in the plane can be similarly shown to have a congruent triangle on the other side of E, so that E is a twofold rotocenter E_2. Since none of A_2, B_2, and D_2 are congruent with each other, E_2 is congruent with none of them: for instance, if E_2 were congruent with D_2, then B_2 and A_2 would necessarily be congruent. Therefore A_2, B_2, D_2 imply *four* twofold rotosimplexes. This is the only example where there are more than three rotosimplexes in the same plane.

These four simplexes constitute the nodes whose meshes have the shape of a general parallelogram (Fig. 23). This shape is dictated by the requirement that every point in the plane must be congruent to a point belonging to any pair of adjacent meshes. Any subdivision of such a parallelogram would not meet this requirement, and hence would not correspond to the definition of a mesh already given. Mesh boundaries for the 2222 net join any rotocenter to rotocenters belonging to *two* of the three rotosimplexes to which it does not itself belong. Just which two are chosen is arbitrary as long as the mesh is a general parallelogram. We shall see in our discussion of enantiomorphy that mirror symmetry will impose restrictions on the shape of the mesh. In that case, the choice of shape of the mesh is no longer arbitrary.

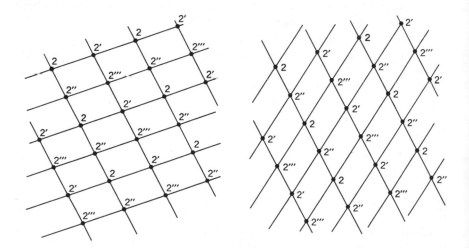

Fig. 23.   Two nets corresponding to four twofold rotosimplexes: the rotosimplexes are in identical configurations, but the meshes are drawn differently.

D.   The solution $1\infty\infty$ and D_p

The solution $1\infty\infty$ corresponds to patterns having translation symmetry in two nonparallel directions and no finite rotation symmetry. When a rotocenter D_p is added to such a pattern, a lattice of rotocenters congruent to D_p is formed. Figure 24 shows two of these rotocenters, D_p and D'_p. If $p > 2$, these two rotocenters interact to generate a sequence of line segments whose intersections are equally spaced along the circumference of a circle. The postulate of closest approach requires that this sequence line closes upon itself after a single revolution, with the result that a rotocenter is implied at the center. This implied rotocenter interacts with D_p according to Theorem 1, so that one of the nets 236, 244, or 333 is generated. The translational symmetries of these nets are visible in Figs. 16, 17, and 18. They are not arbitrary, but limited in directions by the diophantine equation.

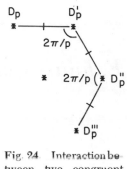

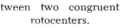

Fig. 24   Interaction between two congruent rotocenters.

If $p = 2$, the two translations operate on D_2 to create the 2222 net already generated in Section 4.C. Thus no new patterns are created from the solution $1\infty\infty$.

E.   Summary

Table 2 summarizes the permitted combinations of rotocenters.

Table 2   Permitted combinations of rotocenters

Number of distinct rotocenters with finite symmetry value	Illustrated by	Combinations of symmetry values
0	The page you are reading	1 (no symmetry)
	Figure 8	∞ (translation symmetry in a single direction)
	Figure 20	$1\infty\infty$ (two nonparallel directions of translational symmetry)
1	Figure 7b	k (any integer $1 < k < \infty$)
2	Figure 12b	22∞
3	Figures 15, 18	236
	Figure 17	244
	Figure 16	333
4	Figure 23	2222

5

Exhaustiveness

A.   Introduction

Although the term "exhaustive" is a technical one, the reader might assume that it reflects the hard labor invested in transgressing the past four chapters. The present chapter will yield some fruits of this investment. We have seen that by systematically building up assemblies of rotocenters, we have eliminated all but a finite number of combinations of rotocenters. Table 2 exhausts all possible types of patterns having rotational symmetry. We see that the wide variety of plane patterns reduces to only nine basic types, each of which has been illustrated by at least one representative example.

The enumeration of Table 2 is called *exhaustive*: no other combinations of rotocenters exist. It must be realized, however, that this table is exhaustive only under the operation of *rotation*; later we shall subdivide each of the nine types further according to the reflection symmetry they may have. Later again, we shall adjoin color symmetry to further extend the variety of types of patterns. Conceivably, additional symmetry operations could be added. It is important, therefore, to specify the operations when exhaustiveness is being discussed.

There are two ways in which exhaustiveness may be approached. One is to enumerate all the patterns one can think of, and then to satisfy oneself that there are no others, either through a formal proof or by comparison with enumerations made by others. This is the course that has been followed in crystallography.† A formal proof, based on a theorem of topology, for the exhaustiveness of plane symmetry patterns, was given by Steiger.‡

The approach followed here is fundamentally different. It consists of tracing the logical implications of, as well as the restrictions on, the coexistence of symmetry elements. Patterns are generated from a small number of coexisting elements, of which all possible combinations are evaluated systematically. The growth of a complex system from a small core that contains all the necessary information is called an *algorismic process*, where the word "algorism" is used in the sense of a generating procedure. Algorisms are used extensively in computer science and in biological science. The procedure we use here for generating symmetry patterns is, in principle, not very different from that used by nature in establishing a genetic code.

The restrictions on the coexistence of rotocenters in a plane as well as the implications thereof can be summarized in a chain of theorems. The first two of these were stated explicitly in Chapter 4; others are implicit in the systematic generation-and-elimination process used here. We shall now state these theorems explicitly with proofs where necessary.

B. A chain of theorems

Theorem 3. The coexistence in a plane of two noncongruent threefold rotocenters implies the existence in that plane of three threefold rotosimplexes. Any straight line joining two noncongruent threefold rotocenters must pass through a threefold rotocenter that is not congruent with either of these two rotocenters.

Proof of Theorem 3. The first half of this theorem follows directly from Theorem 1 and the 333 solution of the diophantine equation. The second sentence is the result of the configuration of the 333 net (Fig. 16); a formal proof does not seem necessary here.

Theorem 4. The existence in a plane periodic pattern of a fourfold rotocenter implies the existence of two fourfold rotosimplexes as well as of a twofold rotosimplex.

†C. Jordan, *Annali di Matematica da Brioschi e Cremona*, Ser. II, TII (1869); L. Sohncke, *Borchard's Journal fuer reine und angew. Mathematik*, **77** (1874), 47–102; F. Klein and R. Fricke, *Vorles, ue. d. Theorie d. ellipt. Modulfunctionen*, Bd. I, Leipzig: Teubner (1890); E. S. Federov, *Zapiski Miner. Imper. S. Petersburg Obshch.*, **28** (1891), 345–390 and tables; A. Schoenflies, *Theorie der Kristallstruktur*, Berlin: Gebr. Borntraeger (1891. Second Edition, 1923); R. Fricke and F. Klein, *Vorles, ue. d. Theorie der automorphen Functionen*, Bd. I. Leipzig: Teubner (1897); A. Speiser, *Die Theorie der Gruppen von endlicher Ordnung*, Ch. VI. Berlin: Julius Springer Verlag (1922; reprint of third edition by Dover Publ., New York, 1943); G. Polya, *Zeit. f. Krist.*, **60** (1924), 278–282.

‡F. Steiger, *Comm. Math. Helv.*, **8** (1936), 235–249.

Proof of Theorem 4. Since the pattern is *periodic*, there must be at least two fourfold rotocenters. In Chapter 4 (see, for instance, Fig. 24) it was shown that a third rotocenter is implied. This implied roto-center may be congruent to the first two, but in that case the argument could be repeated. The postulate of closest approach forbids *indefinite* repetition of the argument, however, so that eventually a rotocenter distinct from the first pair would be implied. According to Theorem 1, its symmetry value would be restricted by the diophantine equation. Only one solution of this equation corresponds to fourfold symmetry, namely, 244, q.e.d.

Theorem 5. The existence in a plane periodic pattern of a sixfold roto-center implies the existence of a twofold, a threefold, and a sixfold rotosimplex.

Proof of Theorem 5. The proof of this theorem is entirely analogous to that of Theorem 4.

Theorem 6. In a plane, the rotosimplex having the highest finite sym-metry value constitutes a lattice.

Proof of Theorem 6. It is recalled that a lattice is an array of points related to each other by translation symmetry. The environments of points belonging to the same lattice not only must be identical, but they also must have identical orientation. In the cases 22∞, 333, and 2222, all finite rotocenters have the same value. Since rotation through 180° of the environment of a twofold rotocenter actually leaves that environment unchanged, all congruent rotocenters in the 2222 and 22∞ nets have their environments oriented parallel (see Figs. 12*b* and 23), hence belong to the same lattice. Similarly, all rotations through 120° leave the environment of a threefold rotocenter unaffected, so that all congruent rotocenters in the 333-net belong to the same lattice (Fig. 16).

In the case 244, the fourfold rotocenters are unaffected by rotations through 90° or 180°; therefore the environments of congruent four-fold rotocenters are oriented parallel (Fig. 17), and congruent four-fold rotocenters belong to the same lattice. It should be noted that twofold rotocenters are not insensitive to 90° rotations: the twofold rotocenters belong to two different lattices, whose orientations are 90° apart (Fig. 17).

In the case 236, the sixfold rotocenters have their environments unaffected by rotations through either 180° or 120°, hence they belong to the same lattice (Figs. 15 and 18). The threefold rotocenters are turned antiparallel to themselves by a 60° rotation (Figs. 15 and 18), and the twofold rotocenters are also affected by a 120° rotation. These considerations at once lead to Theorem 7.

Theorem 7. When k-fold, l-fold, and m-fold rotocenters coexist in a plane, $k \leq l \leq m$ and m finite, the k-fold rotocenters have their environments oriented in m/k different directions, and the l-fold rotocenters have their environments oriented in m/l different directions.

Theorem 8. Two sixfold rotocenters in a plane necessarily belong to the same lattice.

Proof of Theorem 8. Two sixfold rotocenters imply a periodic pattern. Since, except for the case 2222, not more than three roto-simplexes can coexist in a plane, all sixfold rotocenters in a plane must belong to the same rotosimplex, the other two rotosimplexes being necessarily twofold and threefold (Theorem 5). According to Theorem 6, this rotosimplex constitutes a lattice.

Theorem 9. A threefold and a fourfold rotocenter cannot coexist in a plane.

Proof of Theorem 9. In Chapter 2 it was shown that this combination violates the postulate of closest approach. The theorem also follows directly from the fact that no combination 3,4 exists among the solutions of the diophantine equation.

Theorem 10. The existence of a fivefold rotocenter in a plane precludes the existence of any other rotocenter in the same plane.

Proof of Theorem 10. The existence of a second rotocenter would be restricted by Theorem 1 and the diophantine equation. No solution of the diophantine equation includes the value 5, so that the theorem is proven.

Theorem 11. The existence in a plane of a rotocenter with a finite symmetry value greater than six precludes the existence of any other rotocenter in that plane.

Proof of Theorem 11. Again the existence of a second rotocenter would be restricted by Theorem 1 and the diophantine equation, whose solutions do not include a finite value greater than six.

Theorem 12. The point bisecting a line segment joining two evenfold rotocenters belonging to the same lattice coincides with an evenfold rotocenter.

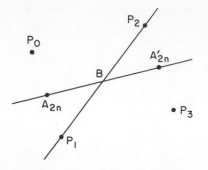

Fig. 25. Two evenfold rotocenters
belonging to the same lattice imply
an evenfold rotocenter.

Proof of Theorem 12. In Fig. 25, A_{2n} and A'_{2n} are the evenfold roto-centers, and B is the point bisecting the line segment $A_{2n}A'_{2n}$. P_0 is an arbitrarily chosen point; the rotational symmetry of A_{2n} implies P_1, congruent with P_0. Since A_{2n} and A'_{2n} belong to the same lattice, P_2 and P_3 are implied, being congruent with P_0 and P_1. Because of these equivalences, $\triangle A_{2n}BP_0$ is congruent with $\triangle A'_{2n}BP_3$. Since P_0 was chosen arbitrarily, this proves that B is an evenfold rotocenter.

Corollary. The rotocenter may be, but generally is not congruent with A_{2n}. If B and A_{2n} belong to the same lattice, then Theorem 12 implies an evenfold rotocenter halfway between A_{2n} and B. This implied rotocenter in turn may or may not belong to the same lattice as A_{2n}; if it does, then yet another evenfold rotocenter is implied. The postulate of closest approach does not permit infinite repetition of this implication, however, so that eventually an evenfold rotocenter is generated on the line segment A_{2n} that does not belong to the same lattice as A_{2n}.

6

Enantiomorphy

A. Enantiomorphic points and rotocenters

In the first chapter we discussed enantiomorphy and congruence of equivalent points. These properties are not mutually exclusive. In an exceptional circumstance two points may be simultaneously enantiomorphic and congruent, as illustrated in Fig. 26, where the points P and P' are both congruent and enantiomorphic. Two points, Q and R, are chosen arbitrarily to define the environment of P. Because of the congruence of P and P', two corresponding points, Q' and R', exist such that $\triangle PQR$ is congruent with $\triangle P'Q'R'$. Because of the enantiomorphy of P and P', two additional points, $\hat{Q}'$ and $\hat{R}'$, exist such that $\triangle PQR$ is congruent with $\triangle P'\hat{Q}'\hat{R}'$. In addition, the congruence of P and P' implies the existence of $\hat{Q}$ and $\hat{R}$ such that $\triangle P'\hat{Q}'\hat{R}'$ is congruent with $\triangle P\hat{Q}\hat{R}$. Therefore $\triangle PQR$ is congruent with $\triangle P\hat{Q}\hat{R}$; similarly, $\triangle P'Q'R'$ and $\triangle P'\hat{Q}'\hat{R}'$ are congruent. Since Q and R were chosen arbitrarily, this means that both P and P' necessarily lie on mirror lines. This leads to Theorem 13.

Theorem 13. Two points that are both congruent and enantiomorphic necessarily lie on mirror lines.

Theorem 13 refers in particular to rotocenters. Two enantiomorphic rotocenters have the same symmetry number. If all rotocenters having the same symmetry number are congruent, e.g., all rotocenters in the 236 system and all twofold rotocenters in the 244 system, then enantiomorphy between such rotocenters implies that they lie on mirrors (Theorem 13); this implies Theorems 14 and 15.

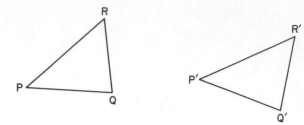

P and P' are congruent

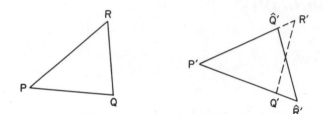

P and P' are enantiomorphic

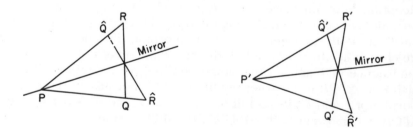

P and P' are congruent as well as enantiomorphic

Fig. 26. Two points being simultaneously congruent and enantiomorphic.

Theorem 14. In a pattern that possesses enantiomorphy every roto-center that has no noncongruent enantiomorph necessarily lies on a mirror line. In particular, any sixfold rotocenter in a pattern that possesses enantiomorphy must lie on a mirror line.

Theorem 15. In a pattern that possesses enantiomorphy, every roto-center not located on a mirror line is limited to the symmetry values $2, 3, 4$, and ∞, and possesses a noncongruent enantiomorph.

Consider two k-fold, enantiomorphic rotocenters A_k and $\hat{A}_k$ (Fig. 27). Choose an arbitrary point P_0: because of the rotocenter A_k, it

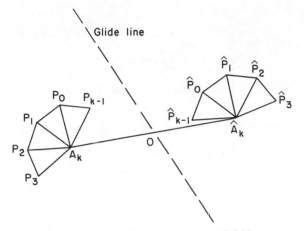

Fig. 27. Two enantiomorphic noncongruent k-fold rotocenters.

Implies k points $P_0 \ldots P_{k-1}$ equidistant from A_k. Enantiomorphy implies k corresponding points $\hat{P}_0 \ldots \hat{P}_{k-1}$ equidistant from $\hat{A}_k$. By definition, a line bisecting the line segment $\hat{A}_k A_k$ and the angle between $A_k P_0$ and $\hat{A}_k \hat{P}_0$ is a *glide line*. This glide line bisects the angle between any pair of lines $A_k P_i$ and $\hat{A}_k \hat{P}_i$, where i is an integer. Since $A_k P_0$ and $\hat{A}_k \hat{P}_1$ are also equivalent, there is also a glide line that bisects the angle between $A_k P_0$ and $A_k \hat{P}_1$, and in general between $A_k P_i$ and $\hat{A}_k \hat{P}_{i+1}$. In total, therefore, k reflection lines intersect at the point O halfway between A_k and $\hat{A}_k$. In general, these are all glide lines, but if one of them perpendicularly bisects $A_k \hat{A}_k$, it is a mirror.

Theorem 16. Two enantiomorphic k-fold rotocenters imply the intersection, at a point halfway between them, of either k glide lines or $(k-1)$ glide lines and a single mirror line.

B.   Coexistence of reflection lines

In Fig. 28, two glide lines, G and G', intersect at an angle ϕ, their respective translation components are $\overline{OP}$ and $\overline{OP'}$. Two successive glide reflections together constitute a direct operation, i.e., a translation or rotation. Therefore a point that is successively glide-reflected in G and in G' is transformed into a point congruent to the original. Peculiar are the intersections of the perpendicular bisectors of line segments $\pm OP$ and $\pm OP'$, marked A, B, C, and D in Fig. 28. Glide reflection of point A in glide line G' produces B, which, by glide reflection in G, in turn produces A again. Therefore point A is invariant

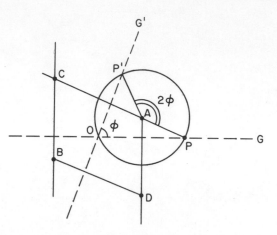

Fig. 28.   Intersecting glide lines.

to successive reflection, i.e., to rotation. The same is true of points B, C, and D. The intersection of glide lines G and G' therefore implies the existence of the four rotocenters A, B, C, and D.

To find the symmetry number of A, let the point P' be glided successively by G' and G. This point is thus turned first into O and then into P; these two successive reflections are equivalent to a rotation around A through an angle 2ϕ. Points P' and P are hence equivalent by rotation so that the symmetry number of rotocenter A equals π/ϕ if ϕ is expressed in radians.

The four rotocenters implied by the coexistence of two glide lines are pairwise enantiomorphic. Only when all four rotocenters are twofold can they all be noncongruent; this may occur when the glide lines are mutually perpendicular. In all other cases not more than three of the rotocenters may be noncongruent, and these are subject to the constraints imposed by the diophantine equation. When we suppose that A and B are not congruent, substituting the relationship $k = \pi/\phi$ into the diophantine equation gives

$$\frac{\phi}{\pi}+\frac{\phi}{\pi}+\frac{1}{m}=1$$

$$\therefore \phi = \frac{\pi}{k} = \left(\frac{m-1}{m}\right)\frac{\pi}{2}$$

Therefore the angle between any two glide lines in a plane is limited to the values 0, 45, 60, and 90°.

If A and B are not congruent, and $\phi \neq 90°$, C and D are necessarily congruent and, according to Theorem 13, must lie on mirror lines. The symmetry numbers are then restricted to the combinations $1\infty\infty$, 244, and 333 (Theorems 15 and 1).

Finally, the possibility remains that A, B, C, and D are all congruent with each other. Such congruency implies an additional rotocenter E, which may or may not be congruent with the first four. If it *is* congruent, then yet another rotocenter is implied, but this implication may not be carried *ad infinitum* because of the postulate of closest approach. Therefore the situation must ultimately arise where two coexisting glide lines imply a triplet of rotosimplexes whose symmetry numbers are constrained by the diophantine equation. Thus we have Theorem 17.

Theorem 17. The acute angle between any two glide lines in a plane is limited to the values 0, 45, 60, and 90°. The coexistence of such glide lines implies the existence of rotosimplexes whose symmetry numbers have the respective combinations $1\infty\infty$, 244, 333, 2222.

If in Fig. 28 we let $\overline{OP}$ go to zero, the glide line G becomes a mirror line. The lines AD and BC then coincide and pass through the point O; two enantiomorphic rotocenters, A and B, are implied. This leads to Theorem 18.

Theorem 18. The acute angle between a mirror line and a glide line in a plane is limited to the values 0, 45, 60, and 90°. The coexistence of such reflection lines implies the existence of rotosimplexes whose symmetry numbers have the respective combinations $1\infty\infty$, 244, 333, 22∞. The line segment joining enantiomorphically paired rotocenters is perpendicularly bisected by the mirror line at the point of intersection of the mirror and glide lines.

Note that a mirror and glide line intersecting perpendicularly imply a *pair* of twofold rotocenters, hence a *linear* array of twofold rotocenters, whereas two mutually perpendicular glide lines imply *four*, and hence a two-dimensional array of twofold rotocenters.

Finally, there is the intersection of two mirror lines. In Fig. 28 both $\overline{OP}$ and $\overline{OP'}$ now vanish and the implied rotocenters coalesce into a single implied rotocenter at the point of intersection O. In Fig. 29 there are two mirrors, M at orientation $\theta = 0$ and M' at orientation $\theta = \phi$. When all the points of M are reflected into M', a mirror line $\hat{M}$ results at orientation $\theta = 2\phi$. Conversely, M' reflected into M produces $\hat{M}'$ at orientation $\theta = -\phi$. Mutual reflections, when continued, produce mirror lines equivalent to M at orientations $\theta = 2n\phi$, n being

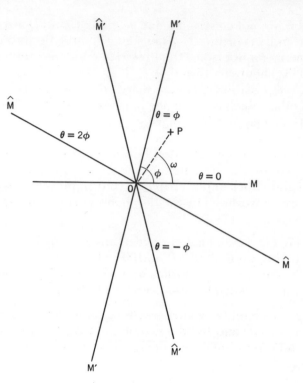

Fig. 29. Intersecting mirror lines.

an integer, and mirrors equivalent to M' at $\theta = (2n+1)\phi$. Since mirror lines extend on both sides of the point of intersection, a mirror equivalent to M at $\theta = 2n\phi$ also exists at $\theta = 2n\phi + \pi$. It is entirely possible that a mirror equivalent to M' also exists at location $\theta = 2n\phi + \pi$; in this case all mirrors are equivalent to each other. This occurs if an integer n' exists so that

$$2n\phi + \pi = (2n'+1)\phi,$$

i.e., when

$$\frac{\pi}{\phi} = 2(n'-n) + 1.$$

Therefore, when π/ϕ is an odd integer, all mirrors are equivalent, but if π/ϕ is even, the mirror lines belong to two distinct sets.

An arbitrary point P at angular position $\theta = \omega$ has a mirror image in mirror M at angular position $\theta = -\omega$, and an image in M' at angular position $\theta = 2\phi - \omega$. These two images are congruent and separated by an angle 2ϕ. There are, through successive multiple reflections, π/ϕ points congruent with P at an equal distance from O; hence O is a center of π/ϕ-fold rotational symmetry. These results are summarized in Theorem 19.

Theorem 19. If two mirror lines intersect at an angle $\theta = j\pi/k$, where j and k are relatively prime integers, $2j \leqslant k$, then a k-fold center of rotational symmetry is implied at the intersection of the mirrors. There are k mirrors intersecting at the center of symmetry. If k is odd, these mirrors are all equivalent, but if k is even, there are generally two distinct sets of mirrors, each containing $\frac{1}{2}k$ mirrors.

If the angle between the mirrors does not meet the specifications of Theorem 19 but π/ϕ is rational, then a pair of mirrors will always be generated by a multiple reflection that does meet these specifications, so that Theorem 19 can be applied to the new pair.

C.   Coexistence of a rotocenter and a reflection line

A rotocenter located on a glide line implies its enantiomorph on the same glide line; the distance between the two enantiomorphs equals the translation component of the glide line. Since two enantiomorphs are not generally congruent, they then do not belong to the same roto-simplex. The coexistence of two noncongruent enantiomorphic roto-centers is therefore subject to the constraints of the diophantine equation; such rotocenters are therefore limited to the symmetry numbers 2, 3, and 4 (Theorem 15). According to Theorem 3, two enan-tiomorphic, noncongruent, threefold rotocenters would be collinear with a third threefold rotocenter distinct from the other two. Since this third rotocenter would also lie on the glide line, it would have a noncongruent image on the same line, thus making *four* noncon-gruent threefold rotocenters. But this is not possible.

Theorem 20. A rotocenter located on a glide line is limited to the sym-metry numbers 2, 4, and ∞, and implies its enantiomorph on the same glide line. The shortest distance between enantiomorphs equals the translation component of the glide line.

Figure 30 shows two evenfold enantiomorphic rotocomplexes, A_{2n} and $\hat{A}_{2n}$, on a glide line. A point P_1 is chosen on the perpendicular bisector of the shortest line segment joining two enantiomorphic roto-centers. The rotocenter A_{2n} implies a point P_2 congruent to P_1. Because of the glide symmetry, P_1 and P_2 are also enantiomorphs. Therefore P_1 and P_2 are congruent enantiomorphs, hence they lie on mirror lines

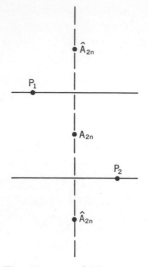

Fig. 30. Evenfold enantiomorphic rotocenters on a
glide line.

(Theorem 13). Since P_1 was chosen anywhere on the perpendicular bisector of a line segment $A_{2n}\hat{A}_{2n}$, such a bisector must itself be a mirror line. Therefore we have a corollary to Theorem 20.

Corollary to Theorem 20. The line segment joining two enantiomorphic rotocenters on a glide line is perpendicularly bisected by a mirror line.

A k-fold rotocenter located on a mirror line may have any integral symmetry value. It generates $k-1$ additional mirror lines, to which Theorem 19 applies, leading to Theorem 21.

Theorem 21. A k-fold rotocenter on a mirror line implies k mirror lines intersecting at equal angles at the center. If k is odd, all mirror lines are equivalent, but if k is even, the mirrors generally belong to two distinct sets.

When k goes to infinity, the distinction between odd and even k made in Theorem 21 becomes meaningless. This case is considered separately and illustrated in Fig. 31. The two parallel mirrors are equivalent. An arbitrary point, P_1, is reflected in M, to produce an enan-

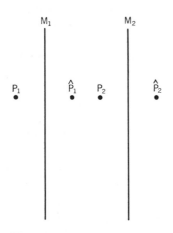

tiomorph $\hat{P}_1$. Because of the equivalence of M_1 and M_2, the pair P_1, $\hat{P}_1$ must be reproduced as $(P_2, \hat{P}_2)$. Between M_1 and M_2 the enantiomorphic pair $\hat{P}_1$, P_2 arises. Since P_1 was chosen arbitrarily, this implies a mirror parallel to and halfway between M_1 and M_2.

Theorem 22. The coexistence in a plane of two equivalent, parallel, mirror line implies a mirror line parallel to and equidistant from these two mirrors.

The implied mirror may be either equivalent to or distinct from the two postulated mirrors. If it is equivalent, then Theorem 22 applies again. Since we are not interested in continuous distribu-

Fig. 31. Two equivalent
parallel mirrors.

tions of mirror lines, however, this chain of implications must be broken after a finite number of times by the implications of a distinct mirror line.

Theorem 23. The existence of a mirror line perpendicular to the direction of translational symmetry implies a set of equidistant mutually parallel, alternatingly distinct mirror lines.

The theorems derived in this chapter give all the constraints on enantiomorphy in the plane. They will provide all the information required for generating a complete set of configurations of symmetry elements in the euclidean plane.

7

Exhaustive generation of enantiomorphic configurations in the plane

A. Adjunction of enantiomorphy to the rotosimplexes

All permitted combinations of rotocenters in the plane were listed in Table 2. With the aid of the theorems derived in Chapter 6 we now adjoin enantiomorphy to the rotational symmetry.

First we consider the cases of a single rotocenter, with both finite and infinite symmetry number. Next are those with three rotosimplexes, in which two possibilities arise. Two rotosimplexes may be enantiomorphically paired, with the third simplex required by Theorem 14 to lie on mirror lines, or all three rotosimplexes may be located on mirror lines. Finally, there are cases of four twofold rotosimplexes in which the four may be pairwise enantiomorphic, or two may be enantiomorphic and the other two distinct from these, or all four may be distinct from each other and all located on mirrors.

In Section 3.D, a fundamental region was defined as a plane-filling region within which no two points are equivalent to each other. The fundamental region generally is not uniquely defined, though its area is. In the absence of enantiomorphy, any two adjacent meshes constitute a possible fundamental region (cf. Fig. 18).

Any point not on a mirror line has an enantiomorph on the opposite side of the mirror. A mirror line therefore always passes halfway between two equivalent points, hence it necessarily constitutes a boundary between fundamental regions. When all rotocenters are located on mirror lines, every mesh is bounded by mirrors and in this case uniquely defines a fundamental region.

When rotocenters are enantiomorphically paired, each such roto-center is generally at the center of a regular polygon bounded by mirror lines. These mirror lines constitute boundaries of fundamental regions; each polygon may further be divided arbitrarily into k congruent fundamental regions meeting at a rotocenter where k equals the symmetry value of the enantiomorphically paired rotocenters.

B. Enantiomorphy with a single finite rotocenter

In Chapter 2 we gave separate consideration to three cases having a single rotocenter: one having symmetry value $k = 1$, one having $1 < k < \infty$, and one having $k = \infty$.

Since there cannot be any congruence when $k = 1$, the existence of a glide line or of more than a single mirror line (Theorem 19) is automatically eliminated. This leaves only a single possibility for enantiomorphy without congruence, namely, the existence of a single mirror (Fig. 32).

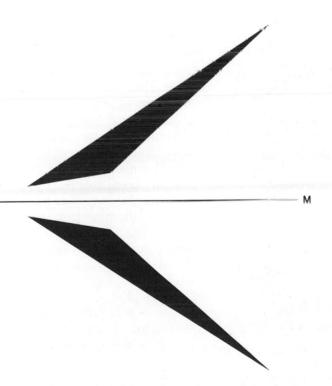
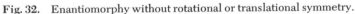

Fig. 32.   Enantiomorphy without rotational or translational symmetry.

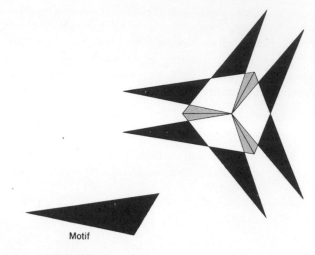

Motif

Fig. 33.   A single rotocenter on a mirror line.

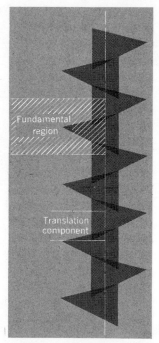

Fundamental region

Translation component

Fig. 34.   Enantiomorphy without finite rotational symmetry: a single glide line.

When $1 < k < \infty$, a glide line is also not permitted as it would imply a second rotocenter. According to Theorem 14, any rotocenter that cannot have a noncongruent enantiomorph must lie on a mirror line. Since here we permit only a single rotocenter, a noncongruent enantiomorph is impossible. Therefore the k-fold rotocenter necessarily lies on a mirror; according to Theorem 21, $k-1$ additional mirrors are then implied. This case is illustrated for $k = 3$ by Fig. 33, which was generated by the successive reflection of an asymmetrical triangle in three mirrors that make equal angles with each other.

When $k = \infty$, the rotation symmetry becomes translation symmetry in a single direction. Theorems 17, 18, and 19 preclude the existence of any reflection lines except a single one or mutually parallel ones. The single reflection line may be a glide line or a mirror line, but in

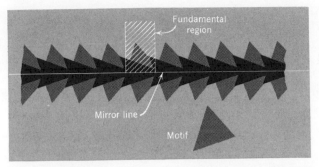
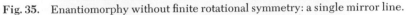

Fig. 35. Enantiomorphy without finite rotational symmetry: a single mirror line.

either case it must be parallel to the direction of translational symmetry. Any other orientation would produce by translation an infinity of reflection lines. These cases are illustrated in Figs. 34 and 35.

Several mutually parallel reflection lines generate translational symmetry in a direction perpendicular to themselves. Glide lines imply translational symmetry parallel to themselves; a glide line parallel to another reflection line therefore implies translation symmetry in two nonparallel directions, hence it is not possible here. This limits us to a set of mutually parallel, alternatingly distinct mirror lines (Theorem 23) as illustrated in Fig. 36.

C.   Enantiomorphy for the 1 ∞∞ system

Since no finite rotational symmetry occurs here, any reflection lines that may occur must be mutually parallel (Theorems 17, 18, and 19).

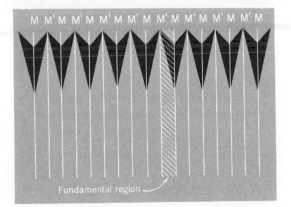

Fig. 36.   Enantiomorphy without finite rotational symmetry: alternating parallel mirrors.

Two parallel reflection lines suffice to generate, by multiple reflection in each other, an infinite array having translational symmetry. Three combinations of such reflection lines exhaust all possibilities: two parallel distinct mirrors, two parallel distinct glide lines, and a glide line parallel to a mirror line. Figures 37, 38, 39 illustrate these three cases, using a sample of wood veneer as fundamental region.

D.   Enantiomorphy for the 22∞ system

In this system all rotocenters are collinear. Any glide line that may exist must pass through the rotocenters, for any rotocenter off a glide line would generate additional rotocenters and result in a 2222 rather than a 22∞ pattern. According to Theorem 20 and its corollary, the resulting pattern consists of a glide line passing through alternating enantiomorphically paired, twofold rotocenters; mirror lines perpendicularly bisect the shortest line segments between enantiomorphic twofold rotocenters (Fig. 40).

Any mirror line necessarily either coincides with or is perpendicular to the line joining the twofold rotocenters. Any other orientation would generate a 2222 array, which will be exhaustively considered later. According to Theorem 21, either orientation implies a mirror in the other orientation, so that there is actually only a single configuration having a twofold rotocenter on a mirror line (Fig. 41). This exhausts the possibilities for enantiomorphy in the 22∞ system.

E.   Enantiomorphy in the 236 system

According to Theorem 14, all rotocenters necessarily lie on mirror lines. Only one configuration (Fig. 42) is therefore possible in this system. Since all congruent rotocenters here are also enantiomorphs, Theorem 16 can be invoked to locate all glide lines (Fig. 42).

F.   Enantiomorphy in the 244 system

Theorem 13 requires all twofold rotocenters to lie on mirror lines, and, according to Theorem 21, such rotocenters lie at the intersection of mutually perpendicular mirror lines. For the fourfold rotocenters there are now two possibilities: they may belong to enantiomorphically paired rotosimplexes, or they may be quite distinct.

In the first case (Fig 43), Theorem 16 and the corollary to Theorem 20 show all mirror lines to pass between enantiomorphically paired fourfold rotocenters (Fig. 43), with glide lines completing the four

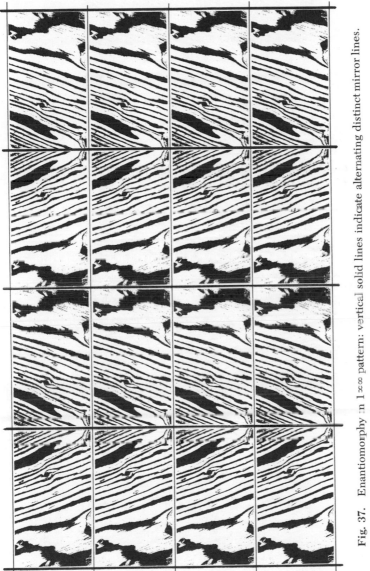

Fig. 37. Enantiomorphy $m\,1\,\infty$ pattern: vertical solid lines indicate alternating distinct mirror lines.

49

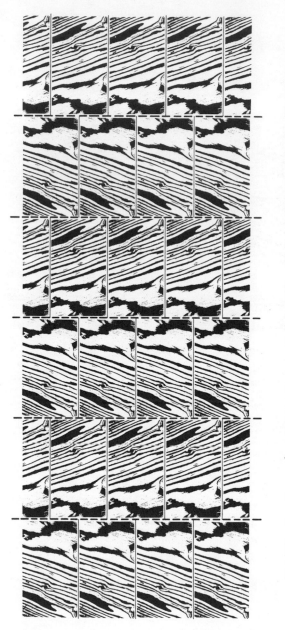

Fig. 38. Enantiomorphy in 1 ∞ pattern: vertical dotted lines indicate alternating distinct glide lines.

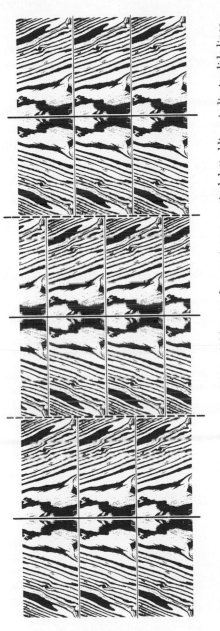

Fig. 39. Enantiomorphy in 1∞∞ pattern: vertical solid lines indicate mirrors, vertical dotted lines indicate glide lines.

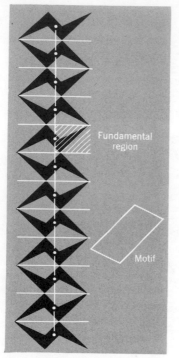

Fig. 40. Enantiomorphy in the
22∞ system: twofold rotocenters
enantiomorphically paired. The
motif used to generate this pattern was a parallelogram shown
explicitly.

required reflection lines intersecting at
equal angles halfway between these
pairs.

In the second case, Theorem 14
requires all fourfold rotocenters to
lie on mirrors. According to Theorem
21, four mirror lines intersect at the
fourfold rotocenters. Application of
Theorem 16 shows that in addition to
the two mirror lines intersecting at
the twofold rotocenters, two glide
lines also intersect at the twofold
rotocenters (Fig. 44).

G.   Enantiomorphy in the 333 system

The threefold rotocenters belong to
three different rotosimplexes. Two of
these may be enantiomorphically paired, but the third must be located on
mirrors (Theorem 14). Again Theorem
16 is invoked to locate all glide lines
(Fig. 45).

Alternatively, all three rotocomplexes may be located on mirrors; according to Theorem 21, three mirror lines

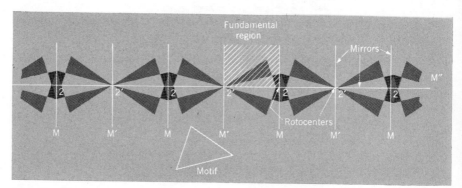

Fig. 41.   Enantiomorphy in the 22∞ system: all rotocenters on mirrors. The motif used
was the triangle shown explicitly.

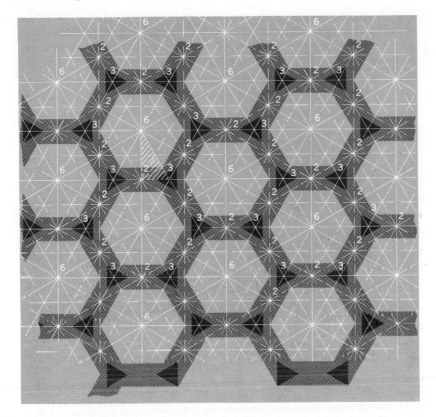

Fig. 42. Enantiomorphy in the 236 system: all rotocenters located on mirror lines.
Solid lines represent mirrors, dashed lines represent glide lines.

then intersect at equal angles at each rotocenter. Theorem 16 is
now applicable to any pair of congruent rotocenters so that the
glide lines can be located (Fig. 46). It should be noted that in this
pattern all triangles are individually congruent, but that their contexts (environments) in the pattern make them distinct. One simplex
of triangles has vertices at the midpoints of edges of the other kind
of triangles, while the latter have free vertices.

It is interesting when comparing Figs. 45 and 46 to note that these
two have identical configurations of mirror and glide lines, but the
locations of rotocenters with respect to these lines are different.

A seemingly natural question is whether a reflection line or a rotocenter is more fundamental. We have already seen that the coexistence of reflection lines implies rotational symmetry, but that rotational
symmetry can exist without reflection symmetry. The examples of

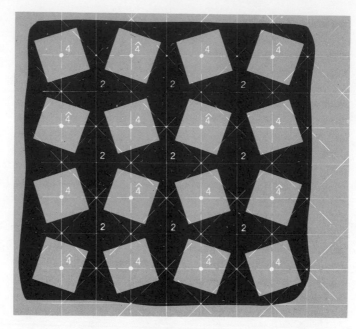

Fig. 43.   Enantiomorphy in the 244 system: noncongruent fourfold rotocenters enantiomorphically paired. $\hat{4}$ indicates enantiomorph of 4.

Figs. 45 and 46 illustrate that the configurations of mirror and glide
lines alone do not uniquely define the symmetry of a pattern, but that
the symmetry relations between rotosimplexes (enantiomorphy or
distinctness) does. This was one of the chief reasons for considering
rotational symmetry as the fundamental starting point for generating
patterns.

H.   Enantiomorphy in the 2222 system

Here, all four rotosimplexes may be pairwise enantiomorphic, or
two may be paired while the others, not possessing enantiomorphs,
lie on mirrors. Finally, all four may lie on mirrors.

Consider first a pair of enantiomorphic twofold rotocenters. According to Theorem 16 two mutually perpendicular reflection lines intersect at a point halfway between the two rotocenters. Generally these
lines are both glide lines (Fig. 47a), but one may be a mirror line
(Fig. 47b). In the former case, a two-dimensional array of twofold
rotocenters is implied by continued reflection in both glide lines;
according to Theorem 12 additional twofold rotocenters are implied,

Fig. 44.   Enantiomorphy in the 244 system: all rotocenters located on mirror lines.

as shown in Fig. 48. The resulting pattern has four pairwise enantiomorphic twofold rotocenters, and is illustrated in Fig. 48.

When one of the reflection lines implied by the enantiomorphic
pair of twofold rotocenters is a mirror line, a 22∞ pattern results. We
may now add to this the second pair of enantiomorphic rotocenters to
generate a 2222 pattern. The location of this second pair is constrained by the fact that the 2222 mesh must be a parallelogram.
Since the mirror perpendicularly bisects the line segment joining the

Fig. 45.   Enantiomorphy in the 333 system: two rotocomplexes are enantiomorphically
paired, the third is located on mirror lines.

first pair of rotosimplexes, the parallelogram here must be a rectangle.
The ratio of edgelengths of this rectangle is totally arbitrary. Figure
49 illustrates this configuration: the four twofold rotocenters generate
an infinity of mutually parallel mirror lines as well as an infinity of

Fig. 46.   Enantiomorphy in the 333 system: all rotocenters lie on mirror lines.

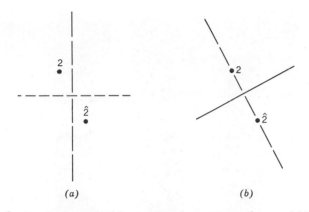

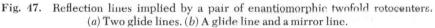

(a)   (b)

Fig. 47.   Reflection lines implied by a pair of enantiomorphic twofold rotocenters.
(a) Two glide lines. (b) A glide line and a mirror line.

glide lines, parallel to each other and perpendicular to the mirror
lines.

Next is the possibility of two enantiomorphically paired rotosimplexes, with the remaining two rotosimplexes distinct from these

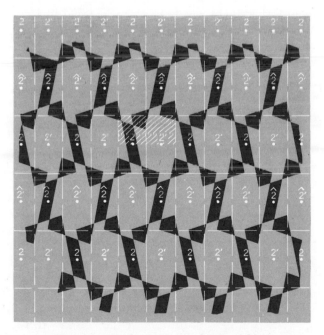

Fig. 48.   Pattern having two pairs of enantiomorphic twofold rotocenters, with mutually parallel and perpendicular glide lines.

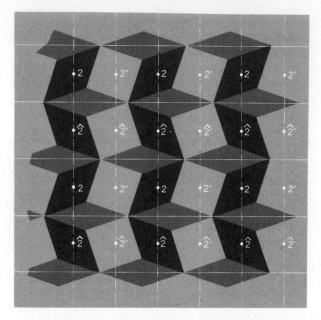

Fig. 49. Pattern having two pairs of enantiomorphic twofold rotocenters: all glide
lines are perpendicular to all mirror lines.

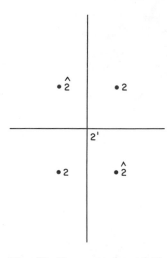

Fig. 50. Two paired twofold
rotocomplexes and a third
complex at the intersection of
two mutually perpendicular
mirror lines.

and from each other. These unpaired
rotocenters lie on mirrors according to
Theorem 14 and, according to Theorem
21, these mirrors intersect perpendicularly at each such rotocenter (Fig. 50).
The coexistence of the three rotosimplexes implies a fourth twofold
rotosimplex, as in Fig. 22. Since this
fourth rotosimplex is not enantiomorphically paired to any other, it
must lie on mirrors (Theorem 14).
The resulting configuration of symmetry
elements is shown in Fig. 51; the glide
lines follow from Theorem 16.

Finally, there is the case where all rotosimplexes are distinct, and hence lie on
mirrors (Theorem 14). In Fig. 52, which
illustrates this case, the motif is a regular
hexagon. To paraphrase a well-known

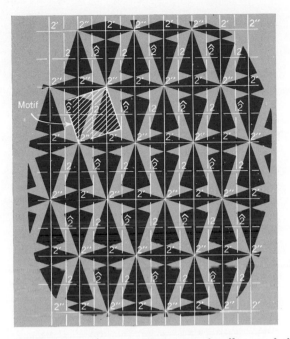

Fig. 51. Two twofold rotocomplexes are enantiomorphically paired, the others lie at the intersections of mutually perpendicular mirror lines.

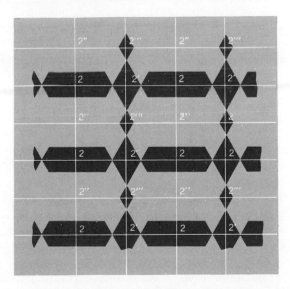

Fig. 52. Four distinct twofold rotocomplexes, all located on mirror lines.

cliche, all that contains a regular hexagon does not have sixfold symmetry. This concludes the exhaustive generation of rotation and reflection symmetry configurations in the plane.

A cautionary note should be inserted here. Theorem 19 states that generally the mirrors intersecting at an evenfold rotocenter belong to two distinct sets. In Fig. 43 the mirrors intersecting at the twofold rotocenters are equivalent, being related by the fourfold rotational symmetry. They are *not* related by the rotocenter at their intersection, for, being twofold, this center could not transform two mutually perpendicular lines into one another. For this reason the qualifier "generally" is used in Theorem 19: special cases of symmetry elements extraneous to the intersecting mirror lines may imply special symmetry relations between the "generally" distinct mirror lines.

A similar situation arises in Fig. 40. It will be noticed that mirror lines that do *not* contain an evenfold rotocenter are *polar*, i.e., no mirror exists perpendicular to them, so that the two opposite directions of the mirror line are distinguishable. Thus we must make a careful distinction between *parallel* and *antiparallel* orientations of such mirror lines: equivalent parallel mirror lines are related by *translational* symmetry, but equivalent *antiparallel* mirror lines are related by twofold rotational symmetry. Theorem 22 speaks of the implication of a mirror line between two *parallel* mirror lines, but is not applicable to *antiparallel* lines, as illustrated in Fig. 40. Here adjacent parallel mirror lines are separated by implied antiparallel mirror lines, and between an adjacent pair of antiparallel mirror lines there is a twofold rotocenter, but *no* mirror line, for Theorem 22 does *not* apply to antiparallel mirror lines.

I.   Summary and nomenclature

The chain of twenty-three theorems accomplishes two purposes: it expresses the restrictions on the coexistence of symmetry elements in the plane, and it establishes the additional symmetry elements implied by the permitted combinations of symmetry elements. Thus, by starting systematically with each permitted combination of symmetry elements, we have generated twenty-eight configurations, of which two have no rotational symmetry, two have a single rotocenter (whose symmetry value may have any finite symmetry number greater than unity), seven are periodic in a single direction, and seventeen are periodic in two different directions. In Table 2 we classified patterns according to their combination of rotocenters. We shall now further classify patterns according to their enantiomorphy.

For the patterns periodic in two directions the international tables of crystallography have a standard nomenclature. Since this nomenclature is not meaningful in the present context, we shall modify it here, listing in Table 3 both the I.T. and our own modified notation.

Table 3   Exhaustive list of configurations of symmetry elements in the plane

Number of distinct rotocenters with finite symmetry value	Combination of symmetry values	Configurations	I.T. notation	Illustrated by Figure number
0	1	1		
		$1m$		32
	∞	∞		8
		$\infty mm'$		36
		∞m		35
		∞g		34
	$1\infty\infty$	$1\infty\infty'$	$p1$	20
		$1\infty\infty'mm'$	pm	37
		$1\infty\infty'mg$	cm	39
		$1\infty\infty'gg'$	pg	38
1	k	k	C_k	7b
		km	D_k	33
2	22∞	$22'\infty$		12h
		$\underline{22'}\infty$		41
		$2\hat{2}\infty$		40
3	236	236	$p6$	15, 18
		$\underline{236}$	$p6m$	42
	244	$244'$	$p4$	17
		$\underline{244'}$	$p4m$	44
		$\underline{2}4\hat{4}$	$p4g$	43
	333	$33'3''$	$p3$	16
		$\underline{33'3''}$	$p3m1$	46
		$3\hat{3}\underline{3}'$	$p31m$	45
4	2222	$22'2''2''''$	$p2$	23
		$\underline{22'2''}2'''$	pmm	52
		$2\hat{2}2'2''$	cmm	51
		$2\hat{2}2'\hat{2}'g/g'$	pgg	48
		$2\hat{2}2'\hat{2}'m/g$	pmg	49

For our notation we use the fact that in most cases the symmetry relation between rotocomplexes completely defines the symmetry of a pattern. For these cases it suffices to indicate distinctness and enantiomorphy; we shall use a prime (′) to denote distinct rotocenters, and a circumflex (ˆ) to denote enantiomorphs. Moreover, we shall underline the symmetry number of a rotosimplex located on mirror lines. Wherever it is necessary to designate the arrangement of reflection lines (e.g., in the case of the two configurations 2$\hat{\underline{2}}$ 2′$\hat{\underline{2}}$′), the use of a diagonal indicates mutually perpendicular reflection lines; two symbols beside each other, e.g., *mg* or *mm*′, indicate parallel lines. Thus all configurations of symmetry elements, including those of Table 2, are listed in Table 3.

A digression into the field of design might be in order here. We have found that patterns may have rotational symmetry only, may have enantiomorphically paired noncongruent rotocenters, or may have all rotocenters located on mirror lines. All these cases are illustrated and indexed in Table 3. Note that these three kinds of patterns have decided characteristics of their own: the patterns without enantiomorphy are wildly dynamic, whereas the patterns whose rotocenters are located on mirror lines are excessively static. The author has found the frameworks with enantiomorphically paired rotocenters the most satisfactory from an esthetic point of view because they give a fine balance between the dynamic and static. One of the drawbacks of the 236 system in design is that no noncongruent enantiomorphic pairs can exist, so that the only choice is between the very dynamic configuration without enantiomorphy and the rather static configuration having all rotocenters on mirrors. A saving grace of the latter is (Theorem 7) the multitude of different orientations in which the twofold and threefold rotocenters find themselves.

In Section 3.D fundamental regions were defined as mutually equivalent regions that together completely cover the plane, and do not contain a pair of equivalent points. In the absence of enantiomorphy the area, but not the shape of a fundamental region is uniquely defined; any pair of adjacent meshes together constitute a fundamental region. (cf. Fig. 83)

When all rotocenters lie on mirror lines, adjacent meshes are enantiomorphs, so that each mesh separately constitutes a fundamental region. Fundamental regions are here bounded by mirror lines; in this case the shape as well as the area of the fundamental region is uniquely defined.

When two *k*-fold rotocenters are enantiomorphically paired, each mesh is bisected by a mirror line; by definition the two enantiomorphic halves of the same mesh cannot belong to the same fundamental region. Each of the paired rotocenters is surrounded by a polygon whose edges are k mirror lines. This polygon encloses k fundamental regions that meet at the central k-fold rotocenters. Each of these fundamental regions is bounded by one of the mirror lines; its other boundary lines can be fixed arbitrarily.

8

Dichromatic symmetry

A. Color

Thus far we have considered patterns whose points are indistinguishable from each other except possibly through the geometrical relationship they bear to each other. Points were distinguished from each other through their *context* (environment) in a pattern, and on this basis any two points are either equivalent to or distinct from each other. We now relax this requirement of absolute equivalence or distinctness; first we illustrate with some examples the reason for such relaxation, and then we address ourselves to such "relaxed symmetry" through some rigid definitions.

As a first example, consider a chessboard: the squares along one diagonal are dark, those along the other diagonal are light in color. At the center of the board there is a center of twofold rotational symmetry, for rotation through 180° leaves the board apparently unaffected. If the board had been rotated 90° instead of 180° around the same center, the effect would be as if the board had remained stationary, but all squares had reversed their colors. The light and dark squares, though geometrically congruent to each other, are distinct according to our definition, for each dark square is surrounded by four light ones and each light square is surrounded by four dark ones.

A similar situation exists in a crystal of sodium chloride; it is a three-dimensional analog of an infinitely extended chessboard. Here each sodium ion is surrounded by six chlorine ions at the vertices of a regular octahedron, and each chlorine ion is surrounded similarly by sodium ions. According to our definitions of symmetry, sodium and chlorine are distinct, and as yet we have no way of indicating the analogy in the environments of sodium and chloride.

In the mineral sphalerite, each sulfur atom is surrounded by zinc atoms at the vertices of a regular tetrahedron, and each zinc atom is surrounded by sulfur atoms in the same manner. If each zinc atom as well as each sulfur atom is replaced by a carbon atom, then with proper scaling the diamond structure results. Since sulfur is distinct from zinc, whereas all carbon atoms in diamond are equivalent, the similarity in coordinations of these atoms in their respective contexts cannot be expressed in terms of our definitions of symmetry.

A third illustration is found in *Grafiek en tekeningen* by M. C. Escher† (Fig. 53). In his introduction to the first edition of this work, Terpstra says that to describe the symmetry of this well-known picture he ignored the color difference between oppositely directed knights because in a strictly mathematical sense these knights are not symmetrically related. MacGillavry‡ takes a more positive view of the same drawing, recognizing the existence of a strictly mathematical operation relating the opposing knights. It is this mathematical operation, relating two geometrically analogous but otherwise distinct objects in a pattern, to which we address ourselves.

The insistence of crystallographers on absolute identity of symmetrically related points is rooted in the x-ray analysis of crystals. Constructive and destructive interference between wavefronts scattered by ions depends on the electronic configurations of the scattering ions being identical. When neutron beams are scattered by ions bearing a magnetic dipole, such ions appear identical to the neutrons only if the orientations of their dipoles are parallel. The same ions that appear identical to x-rays will be distinct from the point of view of neutron beams unless the orientations of their dipoles are parallel.

These examples should suffice to make us relax the requirement of absolute identicality of symmetrically related points. Instead, we define additional symmetry relations between points whose geometrical contexts are analogous, but which are intrinsically different. The intrinsic difference between points, e.g., between sodium and chloride ions in halite, will be symbolized here by different colors. Initially we shall confine ourselves to patterns with two colors only; later we shall proceed to patterns with as many as twelve colors, the upper bound to the number of different colors in a plane within our definition of color symmetry.

† M. C. Escher, *Grafiek en tekeningen.* Tÿl N. V., Zwolle, the Netherlands (1960).

‡ H. MacGillavry, Symmetry Aspects of M. C. Escher's Periodic Drawings. International Union of Crystallography and A. Oosthoek, Utrecht (1965).

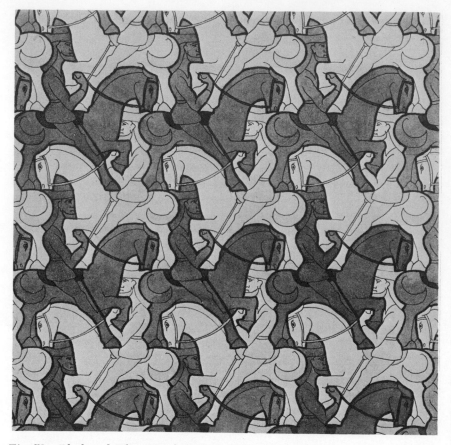

Fig. 53. Black and White Knights, M. C. Escher (reproduced by permission of the artist and the Escher Foundation).

Since two-color patterns are generated within the many-color theory, the separate discussion of dichromatic (two-color) patterns requires some justification here. First, many applications (e.g., in the field of magnetism and neutron diffraction) require only two colors, so that some readers will not want to study the more complex many-color theory. Second, the two-color problem can be treated in a special way that is not just a simple application of the many-color theory. Third, two-color symmetry is unequivocally defined, if only by common usage rather than by precise definition, whereas many-color symmetry will require some arbitrary decisions before we arrive at an exact definition. The difference between the dichromatic and polychromatic treatments given here is like the difference between solving an equation $f(x) = 0$ and plotting a function $f(x)$ for all values of x: in the dichromatic case we know the total number of colors (two), whereas in the polychromatic case we shall generate all patterns, counting the number of colors that results in each instance.

B.   Color reversal

Symmetry was defined in Chapter 1 as an invariance to a coincidence transformation, i.e. a transformation that conserves distance. A definition of color symmetry accordingly requires a transformation involving color. For dichromatic patterns all color transformations involve *color reversal*. We saw previously that a 90° rotation of a chessboard around its center appears equivalent to a reversal of all colors on the board. A slightly different way of looking at this equivalence of the operations of 90° rotation and color reversal on this particular object is to say that the board is invariant to a 90° rotation around its center *followed by* a color reversal. The operation of rotation followed by a color change is called a *color rotation*; an object invariant to color rotation is said to have *color rotational symmetry*. The chessboard thus has not only twofold rotational symmetry, but also *fourfold color-rotational symmetry*.

The same chessboard also has mirror lines; these run diagonally, intersecting perpendicularly at the center. If the board is reflected into a line through its center parallel to its edges, its colors appear also to have been reversed. Thus the board is invariant to reflection into such a line *followed* by color reversal: it has color-reflection symmetry. The board accordingly has fourfold color-rotation symmetry, with four mirror lines intersecting at its center, of which two are "normal" mirrors, and two are color-reversing mirrors.

C.   Normal and reversing symmetry elements

A rotocenter is at the meeting point of several congruent fundamental regions, and a mirror line constitutes the boundary line between enantiomorphic fundamental regions. The colors of points on opposite sides of these elements are generally different; when such points are made to approach, and finally to coincide with the symmetry element, the latter is seen to have an indeterminate color. We therefore never speak of the color of a symmetry element: what matters is whether the symmetry element relates points all of the same or of different colors. A symmetry element that relates points of the same color only is called *normal*; one relating points of two different colors is called *reversing*. All symmetry elements discussed thus far are normal. We shall designate color-reversing symmetry elements by a superscript r and normal symmetry elements either by a superscript n or by the absence of a superscript if clarity is not thereby sacrificed.

D.   Restrictions on the coexistence of the symmetry elements

We shall denote the color of a point by a superscript; in dichromatic patterns we shall use superscripts 0 and 1 only. Any algebraic operation performed on these superscripts will be binary.

First consider a reversing k-fold center of rotational symmetry. If there exists a point P_0^0 of color O, then there are k points (including P_0^0) equivalent to P_0^0 denoted as follows:†

$$P_0^0, \quad P_1^1, \quad P_2^0, \quad P_3^1, \quad P_4^0, \ldots, \quad P_i^{(i \bmod 2)}, \ldots, \quad P_k^{(k \bmod 2)}, \ldots.$$

Here point $P_k^{(k \bmod 2)}$ coincides with P_0^0, $P_{k+1}^{[(k+1)] \bmod 2}$ coincides with P_1^1, etc. The colors of two coincident points must be identical, so that $k_{(\bmod 2)} = 0$, i.e., k must be even. This result leads to Theorem 24.

Theorem 24. Reversing rotocenters necessarily have even symmetry numbers; only evenfold rotocenters may be reversing.

Consider, more generally, a pattern that is invariant to two distinct coincidence transformations, **A** and **B**, both of which may be either normal or reversing. The pattern then has two corresponding symmetry *elements*, which we shall call A and B. The coexistence of two such elements has been shown to imply a third symmetry element, to be called C, to which there corresponds a transformation **C** that is equivalent to successive applications of **A** and **B**. For any point P:

$$CP = B(AP)$$

Therefore, if the transformation **C** changes the color of every point, then either **A** or **B**, but not both, also changes the color of every point. Conversely, if **C** does not cause a color change, then there are two possibilities: either *both* **A** and **B** cause a color change, or *neither* causes a color change.

Theorem 25. If the coexistence of two symmetry elements implies a third symmetry element, the number of reversing elements in this triplet must be even (either zero or two).

†The mod 2 value of an integer is determined by its parity: it equals zero when the integer is even, and equals unity when the integer is odd.

9

Dichromatic configurations without finite rotational symmetry

A. Configurations without congruence

Table 3 (Chapter 7) contains all possible combinations of geometrical symmetry elements. With the aid of Theorems 24 and 25, superscripts r and n can now be assigned to all enumerated configurations of symmetry elements to denote, respectively, reversing and normal elements.

The first configuration in Table 3 is marked 1: there is no geometric symmetry here. Any color transformation in this configuration would merely result in a change in color of every point in the plane. (An illustration would be the transformation of a snapshot negative to the final black-and-white print.) Invariance to such a transformation would be impossible; thus there is no dichromatic 1 configuration. Considering this configuration from the point of view of Theorem 24, one might say that the value $k = 1$, being odd, is forbidden.

Next in Table 3 is the configuration $1m$, corresponding to a single mirror line, which is necessarily dichromatic: every point has an oppositely colored enantiomorph beyond the mirror. This dichromatic configuration shall be called $1m^r$.

B. Configurations with translational symmetry

The configuration ∞ amounts to translation in a single direction; in the sense of Theorem 24 we could not say whether k is even or odd.

However, each translation operation may be accompanied by a color reversal; since the original point is never regenerated by repeated translation, there is no constraint. There is therefore a single dichromatic configuration, labeled ∞^r.

In the configuration $\infty mm'$ any two of the three listed symmetry elements imply the third, so that their coexistence is constrained by Theorem 25. For a dichromatic pattern here we need at least one single reversing symmetry element, and according to Theorem 25 there must be exactly two such elements. There are three ways of distributing two superscripts r over the three symbols: $\infty^r m^r m'$, $\infty^r mm'^r$, and $\infty m^r m'^r$. The first two of these configurations are one and the same, since the assignment of a prime to one of the two sets of mirrors is quite arbitrary. The notations $\infty^r m^r m'$ and $\infty^r mm'^r$ both indicate that one set of mirrors is color reversing and the other is not; the two sets of symbols are therefore degenerate notations for the same configuration. Therefore there actually are only two dichromatic $\infty mm'$ configurations, labeled $\infty^r m^r m'$ and $\infty m^r m'^r$. These are both illustrated in Fig. 54, in which the geometrical reflection is indicated by the shape of the comma, and the color inversion by the color of the body of the comma.

The configuration ∞m corresponds to a single mirror parallel to the translation direction. The translation and reflection operations are independent, so that there is no constraint on the coexistence of the ∞ and m elements. All combinations of reversal and normalcy are therefore possible; there result three dichromatic configurations: $\infty^r m$, ∞m^r, and $\infty^r m^r$ (Fig. 55).

m　　m'ʳ　m　　m'ʳ　m　　m'ʳ　m　　m'ʳ　m　　m'ʳ　m

a) ∞ m m'ʳ : Each unit translation reverses color.

mʳ　m'ʳ　mʳ　m'ʳ　mʳ　m'ʳ　mʳ　m'ʳ　mʳ　m'ʳ　mʳ

b) ∞ mʳ m'ʳ : No color reversal on translation.

Fig. 54.　The dichromatic configurations $\infty mm'$.

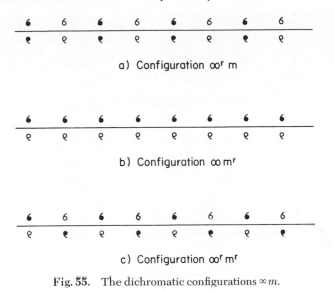

a) Configuration ∞ʳ m

b) Configuration ∞ mʳ

c) Configuration ∞ʳ mʳ

Fig. 55.   The dichromatic configurations ∞ m.

In ∞g the translation and glide operations are not independent, for the unit translation must be commensurate with two successive unit glide operations. Regardless of the character of the glide reflection, the translation will therefore come out normal, so that there can be but a single dichromatic configuration: ∞gʳ (Fig. 56).

Continuing down Table 3, we come to the solution 1∞∞' of the diophantine equation. In the absence of enantiomorphy, the only symmetry is translational in any two nonparallel directions. There is only one dichromatic configuration, illustrated in Fig. 57. Since the translation directions could be chosen independently, this pattern could be interpreted either as the result of one reversing and one normal translation, or as the result of two color reversing translations. In conformity with Theorems 24 and 25, we shall denote this configuration 1∞ʳ∞'ʳ.

The other three configurations in the 1∞∞' system are generated by a pair of parallel reflections lines; in the case 1∞∞'mm' an independent translation parallel to the mirror lines is also required. We shall call this independent translation ∞'; the translation ∞ is the one implied by the parallel mirror lines. The notations mʳm' and mm'ʳ again

Fig. 56.   The dichromatic ∞g configuration.

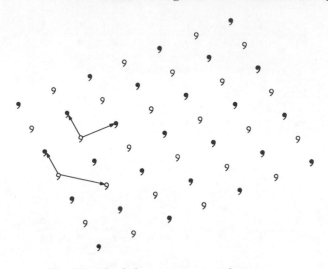

Fig. 57.   The dichromatic $1\infty\infty'$ configuration.

are degenerate symbols for the same configuration. Therefore all possible combinations are as listed in Table 4.

In the configuration $1\infty\infty'mg$, the ∞' symmetry is determined by the glide line and, as in the case of ∞g, must be normal. The ∞ symmetry is determined by m and g both, so that there are three dichromatic $1\infty\infty'mg$ configurations, $1\infty^r\infty'mg^r$, $1\infty\infty'm^rg^r$, and $1\infty^r\infty'm^rg$ (Fig. 59).

The configuration $1\infty\infty'gg'$ likewise has positive ∞' only. Since the configurations gg'^r and g^rg' are equivalent, there are only two dichromatic $1\infty\infty'gg'$ configurations, $1\infty^r\infty'g^rg$ and $1\infty\infty'g^rg'^r$ (Fig. 60).

This exhausts the configurations without finite rotational symmetry.

Table 4   Dichromatic $1\infty\infty'mm'$ configurations

m	m'	∞ (implied by m and m')	∞'	Illustrated in Figure 58
n	r	r	n	a
r	r	n	n	b
n	n	n	r	c
n	r	r	r	d
r	r	n	r	e

Note: n indicates normal, r color reversing elements.

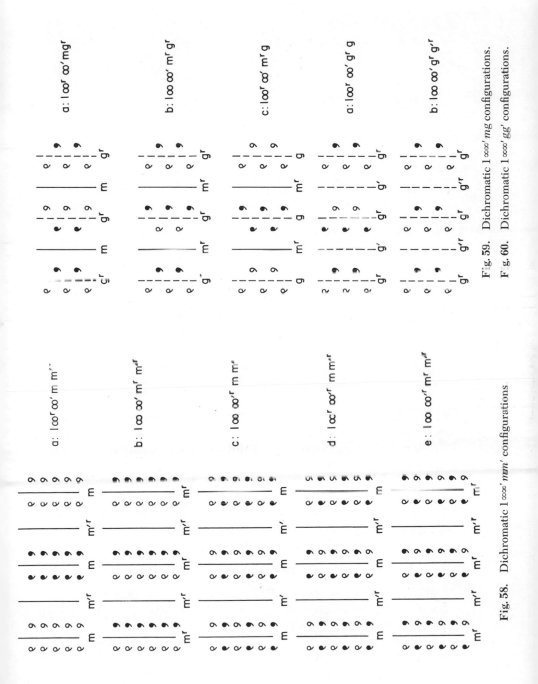

a: l∞ʳ ∞′ mgʳ

b: l∞ ∞′ mʳ gʳ

c: l∞ʳ ∞′ mʳ g

a: l∞ʳ ∞′ gʳ g

b: l∞ ∞′ gʳ gʳ

Fig. 59. Dichromatic l∞∞′ *mg* configurations.

Fig. 60. Dichromatic l∞∞′ *gg*′ configurations.

a: l∞ʳ∞′ m m′⁻

b: l∞ ∞′ mʳ mᵃⁱ

c: l∞ ∞′ʳ m m′⁻

d: l∞ʳ ∞′ʳ m m mʳʳ

e: l∞ ∞′ʳ mʳ m′ᵃʳ

Fig. 58. Dichromatic l∞∞′ *mm*′ configurations

73

10

Dichromatic configurations with finite rotational symmetry

A.   Dichromatic patterns with a single finite rotocenter

According to Theorem 24, the finite rotocenter must have an even symmetry in order to be reversing. The only possible k configuration is therefore $(2\kappa)^r$, where $k = 2\kappa$, and κ is any integer (Fig. 61).

Theorem 21 tells us that a k-fold rotocenter on a mirror line implies k mirror lines intersecting at the rotocenter, which are all equivalent if k is odd but generally belong to two distinct sets if k is even. If k is odd, the center must be normal, and the only possible dichromatic configuration is one in which all mirrors are reversing: km^r (Fig. 62). When k is even, there are two distinct sets of mirrors, which together with the evenfold rotocenter form a triplet in the sense of Theorem 25. Thus there are the following dichromatic configurations: $(2\kappa)m^r m'^r$ (Fig. 63a), and $(2\kappa)^r m^r m'$ (Fig. 63b). These exhaust the dichromatic configurations having a single rotocenter.

B.   Dichromatic patterns with two finite rotocenters

The next configuration in Table 3 is $22'\infty$, which has no enantio-morphy. Theorem 25 permits the following dichromatic configurations: $2^r 2' \infty^r$ (Fig. 64a) and $2^r 2'^r \infty$ (Fig. 64b).

Next consider $\underline{22}'\infty$, which is generated by a single mirror perpendicular to two distinct parallel mirror lines. The twofold rotocenters lie at the intersections of and are implied by the mirror lines. The constraint of Theorem 25 applies therefore. Calling the single mirror m'', the parallel mirrors m and m', we find the following combinations: mm'/m''', mm''/m'', mm''/m''', $m^r m' ^r/m''$, $m^r m' ^r/m'''$. (Note that $m^r m'$ and mm'^r are degenerate notations for the same configurations.) These are all illustrated in Fig. 65.

In configuration $2\hat{2}\infty$ the rotocenters are joined by a glide line; line segments joining nearest rotocenters are perpendicularly bisected by mirror lines (Theorem 18). All colors are determined by the glide line and a mirror line, for such a pair is sufficient to generate the entire configuration. Calling this pair m and g, we find the following dichromatic combinations: mg^r, $m^r g$, and $m^r g^r$ (Fig. 66).

C.   The Dichromatic 236 configurations

According to Theorem 24, the threefold rotocenters are

Fig. 61. Dichromatic pattern $(2\kappa)^r$, illustrated by 6^r.

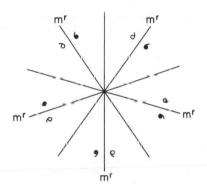

Fig. 62. Dichromatic pattern km^r, illustrated by $5m^r$.

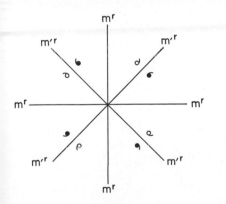

a: $k\,m^r\,m'^r$, illustrated for $k = 4$

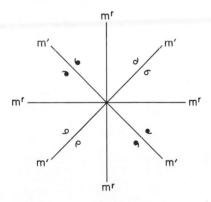

b: $k^r\,m^r\,m'$, illustrated for $k = 4$

Fig. 63.   Dichromatic kmm' configurations.

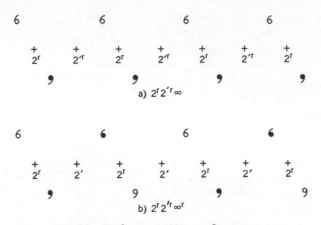

a) $2^r 2'^r \infty$

b) $2^r 2'^r \infty^r$

Fig. 64. Dichromatic $22' \infty$ configurations.

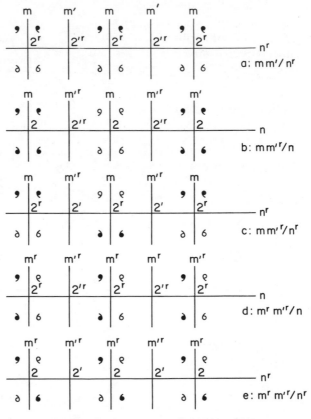

a: $m m' / n^r$

b: $m m'^r / n$

c: $m m'^r / n^r$

d: $m^r m'^r / n$

e: $m^r m'^r / n^r$

Fig. 65. Dichromatic configurations $\underline{22'} \infty$.

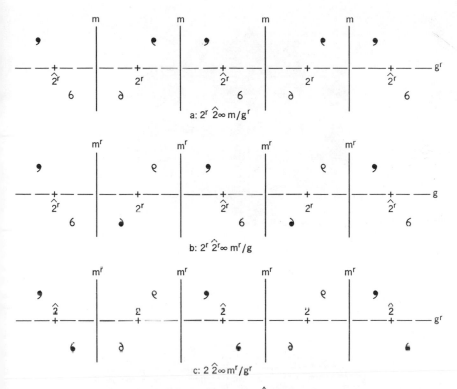

Fig. 66. Dichromatic $2\hat{2}\infty$ patterns.

necessarily normal; since the twofold, threefold, and sixfold rotosimplexes form a triplet in the sense of mutual implication in Theorem
25, the twofold and sixfold rotocenters must both be either normal or
reversing. In the absence of enantiomorphy, dichromaticity is possible only if the twofold and sixfold rotocenters are reversing (Fig. 67).

In $\underline{236}$ configurations all rotocenters lie at the intersections of two
distinct sets of mirror lines; the two kinds of mirrors may be distinguished by the fact that those of one set pass through threefold
centers, whereas the others do not (all pass through the twofold and
sixfold centers). We shall accordingly call the two sets m_{26} and m_{236}.
These mirrors can generate the entire configuration (Theorems 19
and 1) and determine all colors. When all mirrors are normal, all rotocenters are also normal (Theorem 25), so that the resulting pattern is
monochromatic. Therefore for dichromatic patterns there are the
following combinations: $m_{26}{}^r m_{236}{}^r$, $m_{26} m_{236}{}^r$, and $m_{26}{}^r m_{236}$, as illustrated in Fig. 68. This exhausts the dichromatic 236 configurations.

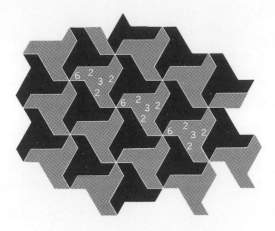

Fig. 67. Dichromatic 236 pattern: $2^r 36^r$.

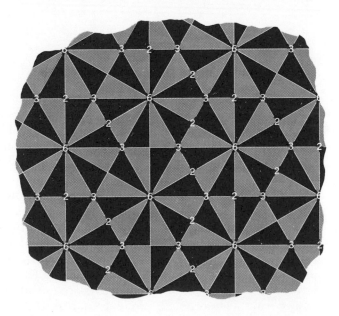

(a)

Fig. 68. (a) Dichromatic $\underline{2}\underline{3}\underline{6}$: $m_{26}^r m_{236}^r$.

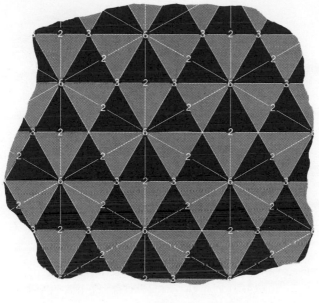

(b)

Fig. 68. (b) Dichromatic $\underline{2}^r\underline{36}^r$: $m_{26}m_{236}^r$.

D. The dichromatic 244′ patterns

First in Table 3 come the configurations without enantiomorphy (244′). According to Theorem 25 two rotosimplexes must be reversing, the third normal; since the notations $4^r4'$ and $44'^r$ are degenerate designations for the same configuration, the following are the only possible combinations: $24^r4'^r$ and $2^r4^r4'$ (Fig. 69). In this illustration there is an "inactive" background color that is unaffected by color rotation. A similar inactive color occurs in Fig. 53, where Escher gives his light as well as his dark knights and horses black eyes.

In the $\underline{244}'$ configurations all rotocenters may be normal, but then dichromaticism is possible only if at least some of the mirrors are color reversing. According to Theorem 25, each rotocenter then must be at the intersection of *two* reversing mirrors. There are three distinct sets of mirror lines (cf. Fig. 44), namely those joining the twofold rotocenters to one simplex of fourfold ones (m_{24}), those joining the twofold centers to the other fourfold simplex ($m_{24'}$), and those joining distinct fourfold rotosimplexes ($m_{44'}$). These mirrors determine all colors. Since the prime was arbitrarily assigned to one rather than the

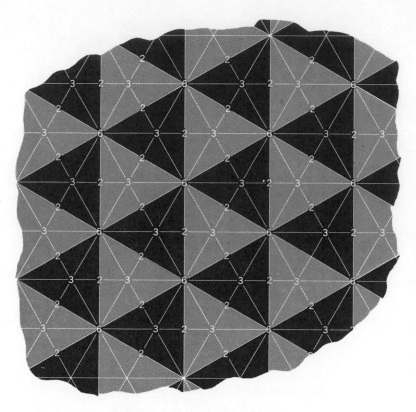

Fig. 68. (c) Dichromatic $\underline{2}^r \underline{36}^r$: $m_{26}^t m_{236}$.

other of the two fourfold rotosimplexes, the notations $m_{24} m_{24'}{}^r$ and $m_{24}{}^r m_{24'}$ are degenerate, leaving the five combinations listed in Table 5. The rotocenters at the mirror intersections are constrained by Theorem 25 and are listed in Table 5, as are the figures in which each configuration is illustrated by a pattern.

Table 5   The dichromatic $\underline{244}'$ configurations

$m_{44'}$	m_{24}	$m_{24'}$	2	4	4'	Illustrated in Figure number
n	n	r	r	n	r	70a
n	r	r	n	r	r	70b
r	n	n	n	r	r	70c
r	n	r	r	r	n	70d
r	r	r	n	n	n	70e

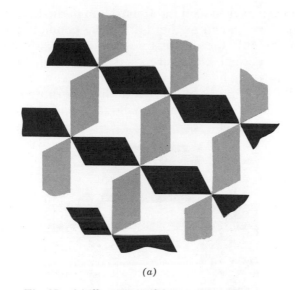

(a)

Fig. 69. (a) Illustration of the configuration $24^r4'^r$.

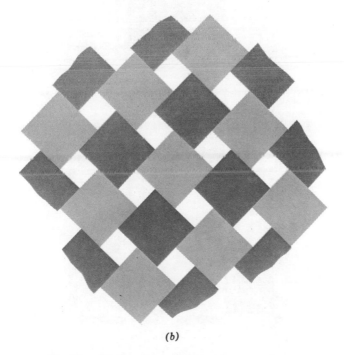

(b)

Fig. 69. (b) Illustration of the configuration $2^r4^r4'$.

81

In Fig. 70 we have reduced the illustration to the barest minimum:
no glide lines are shown explicitly, and no geometrical motifs are
used. Shown are only the mirror lines and centers of rotational symmetry, and only enough colors are used to distinguish geometrically
equivalent points that have different color. Thus for the 244′ patterns
only one color is used per mesh: within one mesh no two points can
be equivalent, and so there is no need to use more than a single color

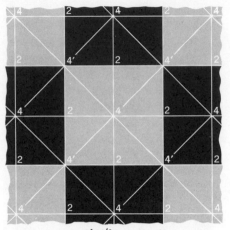

a: $2^r 44'^r\ m_{44'} m_{24}\ m^r_{24'}$

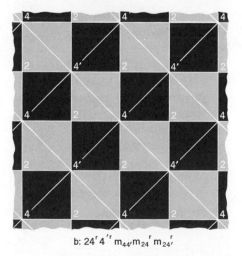

b: $24^r 4'^r\ m_{44'} m_{24}{}^r\ m_{24'}{}^r$

Fig. 70.   Dichromatic 244′ patterns.

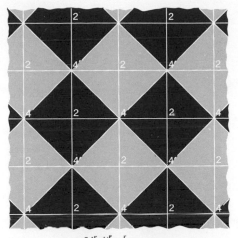

c: 24^r $4'^r$ $m^r_{44'}$ m_{24} $m_{24'}$

d: 2^r 4^r $4'$ $m^r_{44'}$ m_{24} $m^r_{24'}$

Fig. 70.   (cont'd)

for distinguishing points that are already distinct. We shall return to this principle of the most economic use of colors when we discuss polychromatic patterns.

It is interesting to recall that $2k$ meshes meet at a k-fold rotocenter, so that no two points in adjacent meshes may be congruent. Therefore the colors in adjacent meshes are never determined by the rotocenters at which they meet, but only by the sign of the mirror line that

e: 244′ m′$_{44}$′ m$^{c}_{24}$ m$^{c}_{24}$′

Fig. 70. (*cont'd*)

separates them. The colors of next-nearest meshes *are* determined by the rotocenters at which they meet.

Finally, within the 244 system we consider the colors in the $\underline{2}4\hat{4}$ configuration. Since the fourfold rotocenters are enantiomorphically paired, they must have the same sign (Section 8.C). According to Theorem 25, the twofold rotocenters must therefore be normal.

All mirrors are equivalent in the $\underline{2}4\hat{4}$ configuration and intersect at the twofold rotocenters (Fig. 43). A single fourfold rotocenter and a single mirror generate the entire configuration, and define all colors in $\underline{2}4\hat{4}$ patterns. These two elements can occur in combinations $4^r m$, $4m^r$, and $4^r m^r$; the resulting configurations are illustrated in Fig. 71. In the $\underline{2}4\hat{4}$ configuration all meshes are bisected by mirror lines; if the mirror lines are reversing, the two halves must have different colors.

E.   The dichromatic 33′3″ configurations

Since all rotocenters have odd symmetry values, Theorem 24 requires them to be normal. Therefore the only 33′3″ configurations that admit dichromaticity are the enantiomorphic ones.

In $\underline{33}'\underline{3}''$ all rotocenters lie on mirrors; since these mirrors are all equivalent, they must be reversing. There is therefore only a single dichromatic $\underline{33}'\underline{3}''$ configuration (Fig. 72).

In $3\hat{3}'\underline{3}''$ the mirrors are also all equivalent, and must therefore be reversing, generating the sole dichromatic $3\hat{3}\underline{3}'$ configuration (Fig. 73).

These two configurations exhaust the dichromatic 33′3″ configurations.

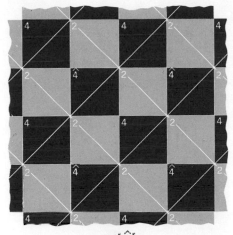

a: $\underline{2}\,4^r\,\widehat{4}^{\,r}m$

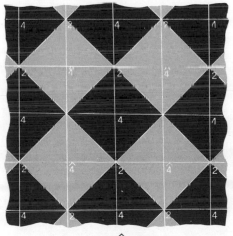

b: $\underline{2}\,4\,\widehat{4}\,m^r$

Fig. 71.   Dichromatic $\underline{2}4\widehat{4}$ patterns.

F.   The dichromatic $22'2''2'''$ configurations

Table 3 lists five of these geometric configurations, the only ones having four noncongruent rotosimplexes. In Fig. 74 there is a mesh in the $22'2''2'''$ net. A point P_0 is arbitrarily chosen; together with rotocenter 2 it implies a point P_1 congruent with P_0. In turn, P_1 and $2'$ together imply P_2, through which $2''$ generates P_3. Finally, $2'''$ relates

c: 2 4^r $\widehat{4}^r$ m^r

Fig. 71.   (cont'd)

P_3 and P_0 by twofold rotational symmetry, closing the circuit. The colors of P_1, P_2, and P_3 are determined by the color of P_0 and the rotocenters 2, 2', and 2'': P_0 and P_3 have the same color if 2''' is normal, but have different colors if 2''' is reversing. This leads to Theorem 26.

Fig. 72.   Dichromatic $\underline{3}\underline{3}'3''$ configuration.

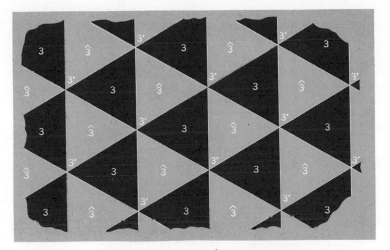

Fig. 73.   Dichromatic 3$\hat{3}$3' configuration.

Theorem 26. Of four coexisting twofold rotosimplexes the number of color-reversing ones must be even (0, 2, or 4).

Thus there are two possibilities for dichromatic configurations without enantiomorphy, namely those with two and those with four reversing rotocenters (Fig. 75). In the case of two reversing rotocenters it makes no difference which rotocenters are reversing, as meshes may be chosen with a certain freedom (Fig. 23).

In the configuration 22'2"2''' (Fig. 76) all rotocenters are located at the intersections of mutually perpendicular mirrors. These mirrors belong to four sets, of which two are mutually parallel, and perpendicular to the other two. These four sets of mirrors, uniquely determine the colors of 22'2" 2''' patterns. The mirrors are designat-

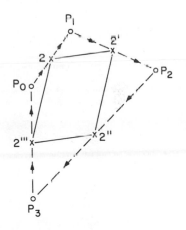

Fig. 74.   Circuit joining four congruent points in a 22'2"2" net.

ed according to the rotocenters which they join: $m_{22'''}$ and $m_{2'm_{2''}}$ are mutually parallel, and perpendicular to $m_{22'}$ and $m_{2''2'''}$. Remembering the fact that primes are assigned arbitrarily, so that there may be degeneracy in nomenclature, we may generate the combinations listed in Table 6.

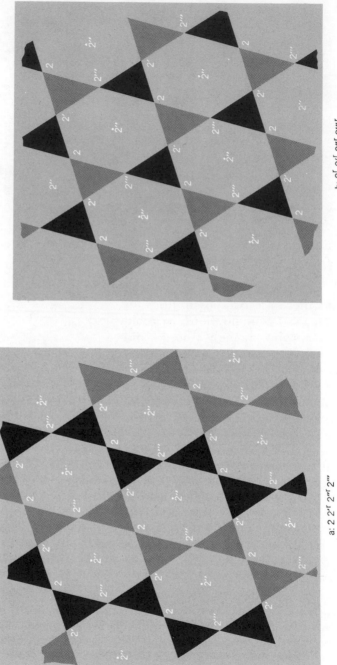

a: $2 \; 2^r \; 2^{rr} \; 2^{rrr}$

b: $2^r \; 2^r \; 2^{rr} \; 2^{rrr}$

Fig. 75. Dichromatic $22'2'2''$ configuration.

88

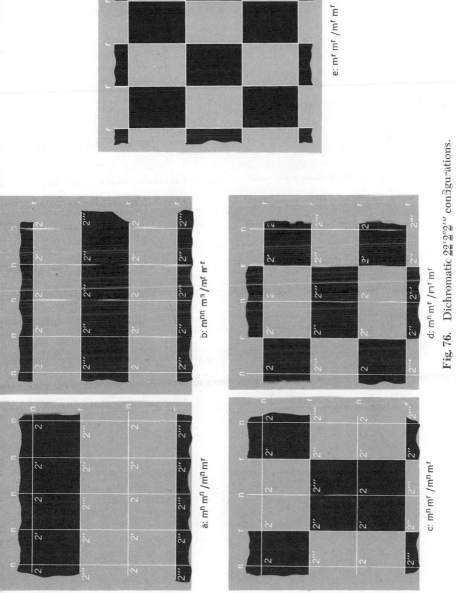

e: $m^r m^r / m^r m^r$

b: $m^n m^n / m^r m^r$

d: $m^n m^r / m^r m^r$

a: $m^n m^n / m^n m^r$

c: $m^n m^r / m^n m^r$

Fig. 76. Dichromatic $2 2' 2' 2'$ configurations.

89

Table 6   Possible combinations in dichromatic $2\underline{2}'2''\underline{2}'''$ configurations

$m_{22'''}$	$m_{2'2''}$	$m_{22'}$	$m_{2''2'''}$	2	2'	2''	2'''	Illustrated in Figure number
n	n	n	r	n	n	r	r	76a
n	n	r	r	r	r	r	r	76b
n	r	n	r	n	r	n	r	76c
n	r	r	r	r	n	n	r	76d
r	r	r	r	n	n	n	n	76e

The configuration $2\hat{\underline{2}}'2''$ (Fig. 77) can be completely generated by two mutually perpendicular mirrors and a single twofold rotocenter not on these mirrors. The single rotocenter, when reflected in either mirror, produces a noncongruent enantiomorph, whereas the mirrors imply a third noncongruent rotocenter at their intersection (Theorem 19). The three noncongruent centers then imply a fourth noncongruent center (Theorems 12 and 2); the four rotocenters generate all additional mirrors, which therefore belong to two sets in this instance. The single twofold rotocenter and two mutually perpendicular mirror lines determine all colors in a $2\hat{\underline{2}}'2''$ pattern. The combinations for this configuration are listed in Table 7.

There remain, in Table 3, the two configurations having two pairs of enantiomorphic rotocenters: $2\hat{\underline{2}}'\hat{2}'g/g'$ and $2\hat{\underline{2}}'\hat{2}'m/g$. The first can be generated by a pair of mutually perpendicular glide lines

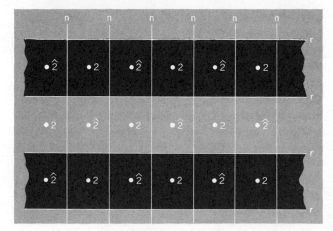

a: $2\,\hat{2}\,\underline{2}'^r\,\underline{2}''^r\,m^n/m^r$

Fig. 77.   Dichromatic $2\hat{\underline{2}}'2''$ configurations.

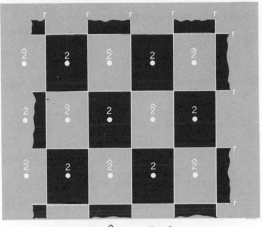

b: $2\ \hat{2}\ \underline{2}'\ \underline{2}''\ m^r/m^r$

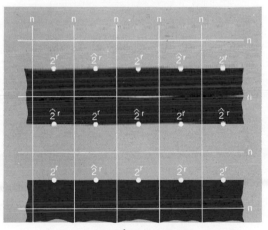

c: $2^r\ \hat{2}^r\ \underline{2}'\ \underline{2}''\ m^n/m^n$

Fig. 77. (cont'd)

(Theorem 17), so that there are just two possibilities: g^r/g'^n (Fig. 78a) and g^r/g'^r (Fig. 78b). The second configuration may be generated by a pair of distinct twofold rotocenters and a mirror line parallel to and not coincident with a line joining these twofold centers. These three elements determine the colors of a $2\hat{2}2'\hat{2}'\,m/g$ pattern; since each rotocenter is paired to an enantiomorph of the same super-script, Theorem 26 is always satisfied. There are five possible combinations:

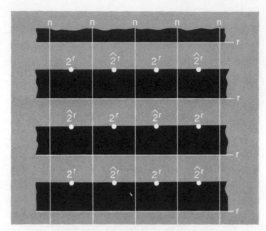

d: $2^r \, \hat{2}^r \, 2'^r \, 2''^r \, m^n / m^r$

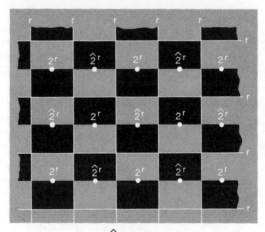

e: $2^r \, \hat{2}^r \, \underline{2}' \, \underline{2}'' \, m^r / m^r$

Fig. 77. (*cont'd*)

Table 7   **Possible combinations in dichromatic $22\underline{2}'2''$ configurations**

m	m'	2	$\hat{2}$	$\underline{2}'$	$\underline{2}''$	Illustrated in Figure number
n	r	n	n	r	r	$77a$
r	r	n	n	n	n	$77b$
n	n	r	r	n	n	$77c$
n	r	r	r	r	r	$77d$
r	r	r	r	n	n	$77e$

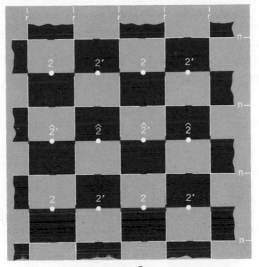

a: g^r/g^n

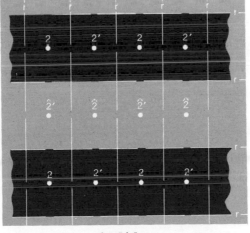

b: g^r/g^r

Fig. 78.   Dichromatic $2\hat{2}2'\hat{2}'g/g'$ configurations.

$22'm^r$, $2^r2'm$, $2^r2'm^r$, $2^r2'^rm$, $2^r2'^rm^r$ (Fig. 79). These exhaust the dichromatic configurations in the plane.

G.   Summary and conclusions

In Chapters 8, 9, and 10 we have relaxed the requirement of absolute identicality between geometrically equivalent points, allowing

a: 2 2′ m^r

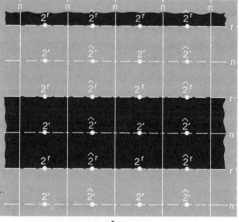

b: 2^r 2′ m

Fig. 79. Dichromatic $2\hat{2}2'\hat{2}'m/g$ configurations.

such points one of two possible attributes which they may or may not share. These attributes are symbolized by color. A symmetry element relating points of the *same* color only is called *normal*, one relating points of *different* colors is called *reversing*.

Accordingly, the principle followed in Chapters 1 to 7 for exhaustively generating all geometrical configurations was followed again.

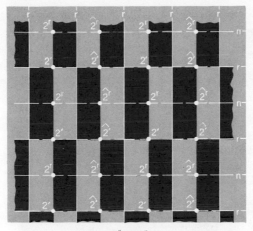

c: $2^r\ 2'\ m^r$

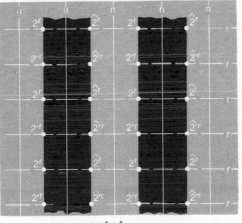

d: $2^r\ 2'^r\ m$

Fig. 79.   (cont'd)

Coexistence of color symmetry elements was found to be subject to constraints and also to imply additional color symmetry elements (Theorems 24, 25, and 26). The resulting dichromatic configurations are listed in Table 8. The second column in Table 8 can stand some explanation, particularly the heading *"May be generated by."* In the derivation of the dichromatic configurations we chose two, three, or (on one occasion) four symmetry elements from which the corresponding geometric configuration may be generated by a chain of

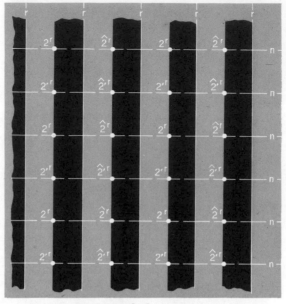

e: $2^r \, 2'^r \, m^r$

Fig. 79.   (*cont'd*)

implications, using any of the Theorems 1 to 24. The choice of these particular elements is to a certain extent arbitrary, though the number necessary to generate any one configuration is characteristic for that configuration; it could be considered as the number of degrees of freedom of that configuration. Since the symmetry elements of a configuration are related by *mutual* implication, as exemplified in Chapter 3, Fig. 15, there are generally several possible combinations of symmetry elements from which the same configuration can be generated. The elements chosen in our derivations appeared to the author to be the most convenient, but the reader is invited to generate the same configurations from different combinations of elements. Because a chain of implications eventually leads to a complete and closed system, the choice of generating elements does not matter, as long as these elements are independent, and their combinations are considered exhaustively.

Of the configurations listed in Table 8, 17 are periodic in a single direction only (those corresponding to the configurations ∞, $\infty mm'$, ∞m, ∞g, $22'\infty$, $\underline{22}'\infty$, and $2\hat{2}\infty$), and 46 are periodic in 2 directions. It is interesting to note the recurrence of certain numbers: 17 dichromatic

Table 8   Summary of dichromatic configurations

Geometric configuration	May be generated by	Dichromatic configuration	Illustrated in Figure number	Shubnikov/Belov notation
1	—	—	—	
$1m$	m	$1m^r$	—	
∞	∞	∞^r		$p'1$
$\infty mm'$	mm'	$\infty^r m^r m'$	$54a$	$p'm1$
		$\infty m^r m'^r$	$54b$	$pm'11$
∞m	∞m	$\infty^r m$	$55a$	$p'1m1$
		∞m^r	$55b$	$p1m'1$
		$\infty^r m^r$	$55c$	$p'1a1$
∞g	g	∞g^r	56	$p1a'1$
$1\infty\infty'$	$\infty\infty'$	$1\infty^r\infty'^r$	57	p'_b1
$1\infty\infty'mm'$	$\infty'mm'$	$1\infty^r\infty'mm'^r$	$58a$	p'_b1m
		$1\infty\infty'm^rm'^r$	$58b$	pm'
		$1\infty\infty'^rmm'$	$58c$	$p'_bm(\equiv g')$
		$1\infty^r\infty'^rmm'^r$	$58d$	$c'm$
		$1\infty\infty'^rm^rm'^r$	$58e$	$p'_bg(\equiv m')$
$1\infty\infty'mg$	mg	$1\infty^r\infty'mg^r$	$59a$	p'_cm
		$1\infty\infty'm^rg^r$	$59b$	cm'
		$1\infty^r\infty'm^rg$	$59c$	p'_cg
$1\infty\infty'gg'$	gg'	$1\infty^r\infty'g^rg'$	$60a$	p'_b1g
		$1\infty\infty'g^rg'^r$	$60b$	pg'
k	k	$(2\kappa)^r$	61	
km	km	km^r	62	
		$(2\kappa)m^rm'^r$	$63a$	
		$(2\kappa)^rm^rm'$	$63b$	
$\underline{22'}\infty$	$22'$	$2^r2'\infty^r$	$64a$	$p'112$
		$2^r2'^r\infty$	$64b$	$p112'$
$\underline{22'}\infty$	mm'/n	$2^r2'^r\infty mm'/m''^r$	$65a$	$pmm'2'$
		$\underline{22'}^r\infty^r mm'^r/m''$	$65b$	$p'mm2$
		$2^r2'\infty^r mm'^r/m''^r$	$65c$	$p'ma2$
		$2^r2'^r\infty m^rm'^r/m''$	$65d$	$pm'm2'$
		$\underline{22'}\infty m^rm'^r/m''^r$	$65e$	$pm'm'2$
$2\hat{2}\infty$	m/g	$2^r\hat{2}^r\infty m/g^r$	$66a$	$pma'2'$
		$2^r\hat{2}^r\infty m^r/g$	$66b$	$pm'a2'$
		$2\hat{2}\infty m^r/g^r$	$66c$	$pm'a'2$
$\underline{236}$	23	2^r36^r	67	$p6'$
$\underline{236}$	$m_{26}m_{236}$	$\underline{236}m^r_{26}m^r_{236}$	$68a$	$p6m'm'$
		$2^r\underline{36}^r m_{26}m^r_{236}$	$68b$	$p6'mm'$
		$2^r\underline{36}^r m^r_{26}m_{236}$	$68c$	$p6'm'm$
$\underline{244'}$	24	$24^r4'^r$	$69a$	$p4'$
		$2^r4^r4'$	$69b$	p'_c4

97

Table 8   Summary of dichromatic configurations (continued)

Geometric configuration	May be generated by	Dichromatic configuration	Illustrated in Figure number	Shubnikov/Belov notation
$\underline{2}44'$	$m_{44'}m_{24}m_{24'}$	$2^r\underline{44}'^r m_{44'}m_{24}m^r_{24'}$	70a	p'_c4mm
		$\underline{24}^r\underline{4}'^r m_{44'}m^r_{24}m^r_{24'}$	70b	$p4'm'm$
		$\underline{24}^r\underline{4}'^r m^r_{44'}m_{24}m_{24'}$	70c	$p4'mm'$
		$2^r\underline{4}^r\underline{4}'m^r_{44'}m_{24}m^r_{24'}$	70d	$p'4gm$
		$\underline{244}'m^r_{44'}m^r_{24}m^r_{24'}$	70e	$p4m'm'$
$\underline{24}\hat{4}$	$4m$	$\underline{24}^r\hat{4}^r m$	71a	$p4'g'm$
		$\underline{244}m^r$	71b	$p4g'm'$
		$\underline{24}^r\hat{4}'^r m^r$	71c	$p4'gm'$
$\overline{33'3''}$	$33'$	—	—	
$\underline{33'}3''$	$3m$	$\underline{33'}3''m^r$	72	$p3m'$
$3\hat{3}3'$	$3m$	$3\hat{3}3'm^r$	73	$p31m'$
$\overline{22'2''2'''}$	$22'2''$	$2^r2'^r2''2'''$	75a	p'_b2
		$2^r2'^r2''^r2'''^r$	75b	$p2'$
$\underline{22'}\underline{2}''2'''$	$m_{22'''}m_{2'2''}/$	$\underline{22'}\underline{2}''2'''^{r\,(a)}$	76a	$p'_b mm$
	$m_{22'}m_{2''2'''}$	$2^r2'^r2''2'''^{r\,(a)}$	76b	pmm'
		$\underline{22'}^r2''2'''^{r\,(a)}$	76c	$c'mm$
		$2^r\underline{2}'2''2'''^{r\,(a)}$	76d	$p'_b gm$
		$\underline{22'}\underline{2}''2'''^{(a)}$	76e	$pm'm'$
$2\hat{2}\underline{2}'2''$	$2m/m'$	$2\hat{2}2'2''^r m/m'^r$	77a	$p'_c mg$
		$2\hat{2}2'\underline{2}''m^r/m'^r$	77b	$cm'm'$
		$2^r\hat{2}2'2''m/m'$	77c	$p'_c mm$
		$2^r\hat{2}2'\underline{2}''m/m'^r$	77d	cmm'
		$2^r\hat{2}2'\underline{2}''m^r/m'^r$	77e	$p'_c gg$
$2\hat{2}2'\hat{2}'g/g'$	g/g'	$2^r\hat{2}2'^r\hat{2}'^r g^r/g'$	78a	pgg'
		$2\hat{2}2'\hat{2}'g^r/g'^r$	78b	$pg'g'$
$2\hat{2}2'\hat{2}'m/g$	$22'm$	$2\hat{2}2'\hat{2}'m^r/g^rg'^r$	79a	$pm'g'$
		$2^r\hat{2}2'\hat{2}' m/g^rg'$	79b	$p'_b mg$
		$2^r\hat{2}2'\hat{2}' m^r/g^rg'$	79c	$p'_b gg$
		$2^r\hat{2}2'^r\hat{2}'^r m/g^rg'^r$	79d	pmg'
		$2^r\hat{2}2'^r\hat{2}'^r m^r/gg'$	79e	$pm'g$

[a]Because of lack of space, the mirrors are not designated explicitly here, but may be found in Table 6.

configurations periodic in a single direction, as well as 17 monochromatic periodic in 2 directions. There are 7 dichromatic configurations having translation symmetry in a single direction only, and also 7 monochromatic configurations periodic in a single direction. The significance of such numerical coincidences appears to lie in the fact that we are dealing with binary variables: patterns may be generated by reflection lines in several combinations of mirrors and glides, or by dichromatic mirrors, either normal or reversing, etc. Thus the same functions of two or three binary variables recur, giving rise to the same number of combinations.

Shubnikov† has considered grey as well as black-white patterns. He considered dichromatic (specifically: black-white) patterns "polar," and he suggested that superposition of black and white will produce grey, i.e., "neutral" patterns. Thus he does not consider black, grey, and white as *three* colors, but rather as *two* colors, one of which is obtained by superposition of the others. Since a reversing transformation turns black into white and white into black, it turns grey (black and white) into grey (white and black). Therefore a grey point can be symmetrically related only to another grey point, whether their symmetry relation is normal or reversing. Hence grey is suitable as a background or inactive color but cannot coexist as an active color with black and white in the same pattern. Thus Shubnikov found black (or white, in any case "polar"), grey ("neutral"), and black-and-white ("polar") patterns, his black and his grey patterns being just the monochromatic ones generated in Chapters 7 to 10. The 7 monochromatic configurations that are periodic in one direction then become, according to Shubnikov's accounting, 7 black (or white) and 7 grey configurations. When to these are added our 17 dichromatic configurations, there result altogether 31 configurations periodic in a single direction. Also, the 17 black and 17 grey monochromatic plus 46 black-and-white configurations yield a total of 80 configurations periodic in two directions. Our results, translated into Shubnikov's accounting system, agree with those found by Shubnikov himself.

The use of grey poses certain logical problems that would wreak havoc in a polychromatic-symmetry theory. A reversing transformation turns black into white, white into black, but grey into itself. There are therefore "pure" colors (black and white) that are turned into each other, and a "mixed" color (grey) that is turned into itself; in other words, two different "kinds" of color that are quite differently affected by the same symmetry elements. In analogy, one might consider polychromatic patterns with "pure" colors blue, red, and yellow, and "mixed" colors purple, green, and orange. An element that relates a blue point to a red one and a red point to a yellow one would then relate a purple point to an orange one, and an orange point to a green one. "Pure" points would never relate to "mixed" ones and vice versa, and they would therefore not coexist. Presumably superposition of three colors would further segregate colors

†A. V. Shubnikov, and N. V. Belov, *Colored symmetry*. Ed: W. T. Holser, (MacMillan, New York, 1964).

into different types which would never coexist in the same pattern. Utter confusion would result, without much purpose, because to every pattern of pure colors there would correspond an analogous pattern of mixed colors. Colors would have to be designated "pure," "mixed," "doubly mixed," etc. To avoid such confusion and lack of system, we confine ourselves to colors of one kind only: there will be consistent rules about color transformations. A set of colored patterns will then be generated, which may be superimposed at will according to any mixing rules desired, just as black and white can be superimposed to form grey. The patterns to be generated in this monograph can be thought of as forming the unit vectors in a multidimensional space, which may be combined in any desired linear combination.

11

Color transformations

A. Color

In Chapters 1 to 7 symmetrical relations between *identical* points were discussed. In Chapter 8 the definition of symmetry was broadened to apply to analogous, though not necessarily identical points. Color was used to distinguish the former from the latter. In Chapters 8 to 10, two colors were used: analogous points could be either identical or distinct.

In magnetic crystals there are usually several directions along which magnetic dipoles orient themselves.[†] All dipoles are analogous, i.e., identical to x-rays, but not identical to neutrons unless their dipoles are mutually parallel. Several colors would therefore be needed to describe the repetition of the locations *and* orientations of these dipoles. We here extend our investigation of the systematic repetition of colors in the plane to as many colors as possible. To do so, we need some quantitative description of the meaning and behavior of color. Belov et al.[‡] have enumerated polychromatic planar symmetry configurations by starting with Fedorov's 230 spatial geometrical configurations, identifying each elevation out of the plane

[†]A. L. Loeb, A binary algebra describing crystal structures with closely packed anions, *Acta Cryst.*, 11 (1958) 469–476.

[‡]N. V. Belov, On the nomenclature of the 80 plane groups in three dimensions, *Kristall.*, 4 (1959) 775–778 [*Sov. Phys. Crystall.*, 4, 730–733]; N. V. Belov and E. N. Belova, Mosaics for 46 plane (Shubnikov) antisymmetry groups and for 15 (Federov) color groups, *Kristall.*, 2 (1957) 21–22 [*Sov. Phys. Crystall.*, 2, 16–18]; A. V. Shubnikov and N. V. Belov, *Colored symmetry.* W. T. Holser, Ed., New York: MacMillan (1964).

with a color. There is, however, no reason to assume that color is homologous with a geometrical coordinate. It is not surprising, therefore, that MacGillavry† found among M. C. Escher's graphic work some example of color symmetry not included in Belov's enumeration. For this reason we shall here generate the polychromatic configurations independently, beginning with some postulates regulating color transformations.

B.   Consistency

As symmetry is invariance to a transformation, any definition of color symmetry requires consideration of a color transformation. A *pure* color transformation is defined as a transformation affecting the color of every point in a pattern, without moving the pattern. Consider, for example, a colored photograph, its negative, and a black-and-white reproduction of the original. Both the transformation of the negative into a color print and the transformation of the color print into a black-and-white reproduction are *pure* color transformations. The important distinction between the two transformations is that in the transformation of a color negative into a color print there is a one-to-one correspondence between the original and transformed colors: every green point becomes red, red becomes green, etc. Thus all color information is retained in the transformation: two points that initially had the same color will acquire the same new color as a result of the transformation. On the other hand, two points that initially had different colors will never become identically colored as a result of the transformation. Because no color information is lost in the transformation, positive can be transformed into negative and vice versa *ad infinitum*.

Once a color print is transformed into a black-and-white reproduction, there is no way in which the colored representation can be recreated from the black-and-white print. Information about color has been lost in the transformation: both red and green might have been transformed to black, with no information available to indicate which had originally been red, which green.

Just as geometrical symmetry deals with the repetition of geometrical motifs in a pattern, color symmetry is concerned with the repetition of color in the pattern. When a color negative is transformed into a positive print, the repetition of colors is not altered,

† C. H. MacGillavry, *Symmetry aspects of M. C. Escher's periodic drawings*, Utrecht: International Union of Crystallography and A. Oosthoek (1965).

whereas in the transformation into a black-and-white reproduction it is. The transformations in which we shall be interested will preserve the repetition of colors. To this purpose we postulate *consistency*:

1. Two identically colored points will, as a result of the transformation, acquire a single new color.
2. Two points originally having different colors necessarily will have different colors after the transformation.

Two patterns that are related by a pure color transformation that also obey these two postulates have identical color repetitions and will be called *equivalent*.

C.   Cyclic color permutations

In Chapter 2 we noted that a k-fold rotocenter is surrounded by k geometrically equivalent points, $P_0, P_1, \ldots, P_{(k-1)}$, on the circumference of a circle centered on the rotocenter. A rotation around this rotocenter through any integral multiple of $2\pi/k$ radians causes a cyclic permutation of those mutually equivalent points. When all points have the same color, the pattern appears invariant to such a rotation.

If some or all of these points have different colors, a rotation of the sort just described becomes evident to the observer as an apparent cyclic permutation of colors. Such a colored pattern is therefore *not* invariant to a simple rotation. Consider, for instance, a single sixfold rotocenter surrounded by six geometrically equivalent points over which three different colors are distributed. Let the colors be red, yellow, and green. Recalling that two geometrically equivalent points have identical environments (cf., for instance, Fig. 46), we shall postulate similarly that color-equivalent points have identical color environments. In our present example this means that the distribution of colors must be in the order red-yellow-green-red-yellow-green, either in clockwise or counterclockwise direction. This pattern is invariant to a rotation through π radians around the rotocenter, but a rotation through $\pi/3$ or $2\pi/3$ radians appears like a cyclic permutation of colors. Any rotation by which this colored pattern appears to be unaffected would have to be accompanied by a cyclical permutation of colors; in our example a red point would become yellow, a yellow one green, a green one red.

We postulate for the reason illustrated in this example that color rotations involve only cyclical color permutations. This postulate, which admittedly restricts our consideration of colored configurations, is predicated on the cyclical nature of geometrical rotation, and on our general principle that equivalent points must have identical environments in the context of the entire pattern. Because translation is here considered as a special case of rotation, we extend this last postulate to color translation as well.

12

Color symmetry

A. Invariance to color transformation

A pattern that has color symmetry is invariant to a color transformation. For instance, a pattern has color reflection symmetry when reflection in a given line followed by an appropriate change of the color of the points in this pattern produces a result that is indistinguishable from the original pattern. A flag that consists of parallel red, white, and blue zones of equal width has ordinary as well as color reflection symmetry: when reflected in a line perpendicular to the zones, the flag is identically reproduced. However, reflection in a line parallel to and equidistant from the red and blue zones needs to be accompanied by a change of every red point to blue and vice versa (white points remain white) before the original flag is reproduced. The line perpendicular to the zones is an ordinary reflection line, that parallel to the zones is a color-reflection line. At the intersection of these two reflection lines there is a twofold color rotocenter: rotation of the flag through 180° about this point followed by a red ↔ blue interchange of colors reproduces the flag exactly.

B. Color congruence

Any point in a pattern that is invariant to a color translation or color rotation is a member of a set of congruent points. Since we have postulated that a color translation or color rotation involves cyclical permutations, the colors of this set of points occur in a definite, enumerable sequence, and therefore may be put in a numerical order, and assigned non-negative integral values.

Consider as an example three points related by a threefold roto-center in a pattern that also has twofold rotational symmetry. Theorems 1 and 2 tell us that this pattern has sixfold rotational symmetry as well. In assigning numerical values to the colors of our three points under consideration we must not forget the twofold and the sixfold symmetries. If we were to assign numerical values 0, 1, and 2 to the colors of our three points, we might find that we needed intermediate numbers for the colors of additional points related to our original three by sixfold or twofold symmetry. We shall therefore start by assigning a general value s to a given point P_0 in a pattern, and a value $s + \Delta s$ to another point P_1 congruent to P_0. The same transformation that turns P_0 into P_1 also turns P_1 into a point P_2 whose color has the value $s + 2\Delta s$, the latter into P_3 having a color of value $s + 3\Delta s$, etc. The total number of colors in a pattern may be finite or infinite. We shall call this total number N. The list of colors in a pattern contains exactly N items, and is interpreted cyclically: the entry following the last entry is again the first one. Thus the numerical value of color is to be interpreted modulo N. (The modulo N value of s is the smallest non-negative remainder after an integral multiple of N has been subtracted. For example, the modulo 7 value of 24 equals 3, the modulo 5 value of 25 equals zero.)

C.   Permutations of colors

Eventually we shall generate all possible colored configurations exhaustively. To do so, it will be necessary to avoid duplications. Although a completely general rule for avoiding duplication has not been found, some particular types of duplication can be prevented by observing special rules.

It is recalled that a consistent color transformation conserves the repetition pattern of colors: only isochromatic pairs remain isochromatic. In our exhaustive enumeration we are interested in all different repetition patterns regardless of the absolute colors. Therefore equivalent patterns count only once; such patterns are related by a pure, consistent color transformation. Patterns whose colors are permuted are equivalent; hence they should be counted once. Two important permutations to watch for are cyclic permutations and reversals (Fig. 80); the three patterns of Fig. 80 are equivalent from the point of view of color symmetry.

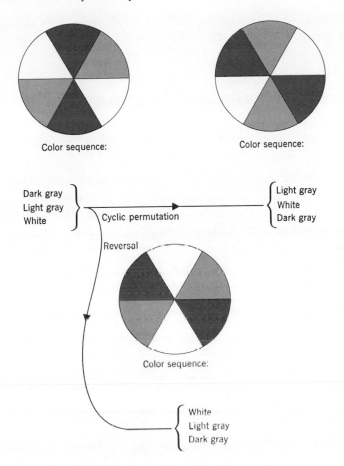

Fig. 80.   Equivalent colored patterns.

D.   Color translational symmetry

We shall henceforth indicate the color of a point by a superscript, remembering that this superscript is interpreted modulo N. A simplex of congruent points is then denoted as follows:

$$P_0^s, \quad P_1^{s+\Delta s}, \ldots, \quad P_j^{s+j\Delta s}, \ldots.$$

In the case of translational symmetry in a single direction only (configuration ∞ in Table 3), the numbers $s, \ldots, s+j\Delta s, \ldots$, increase monotomically until the value N is reached, and then continue from zero upward, repeating periodically after N items if N is finite. The value of N may be chosen arbitrarily.

Similarly, for the configuration $1\infty\infty'$ the points can be denoted by a matrix:

$$
\begin{array}{ccccc}
P_{00}^s, & P_{01}^{s+\Delta s}, & \ldots, & P_{0j}^{s+j\Delta s}, & \ldots \\
P_{10}^{s+\Delta t}, & P_{11}^{s+\Delta s+\Delta t}, & \ldots, & P_{1j}^{s+j\Delta s+\Delta t}, & \ldots \\
\vdots & \vdots & \vdots & \vdots & \vdots \\
P_{i0}^{s+i\Delta t}, & P_{i1}^{s+\Delta s+i\Delta t}, & \ldots, & P_{ij}^{s+j\Delta s+i\Delta t}, & \ldots
\end{array}
$$

The numbers of colors, N, may again be chosen arbitrarily, causing colors to start repeating the same sequence whenever the expression $(s+j\Delta s+i\Delta t)$ reaches the value N. Translational color symmetry does not provide further problems or interest.

E. Colored configurations having a single finite rotocenter

In the presence of a single k-fold rotocenter, there is a complex of exactly k equivalent points:

$$
P_0{}^s, \quad P_1{}^{s+\Delta s}, \ldots, \quad P_j^{s+j\Delta s}, \ldots, \quad P_{k-1}^{s+(k-1)\Delta s}.
$$

The points P_0 and P_k coincide; consistency requires their colors to be identical:

$$
s+k\Delta s \stackrel{N}{=} s, \qquad \text{where } \stackrel{N}{=} \text{indicates identity modulo} -N;
$$
$$
\therefore k\Delta s = pN, \qquad \text{where } p \text{ is an integer}, 0 \leq p < k.
$$

(2)

The color relation between adjacent points, Δs, is thus quantized according to (2); the integer p determines the sequence of colors among equivalent points. This integer will be called the *color parameter* of the k-fold rotocenter. It follows from (2) that the equivalent points have the following colors, listed in proper order:

$$
s, \quad s+\frac{pN}{k}, \quad s+\frac{2pN}{k}, \ldots, \quad s+\frac{jpN}{k}, \ldots, \quad s+\frac{(k-1)pN}{k}.
$$

These numbers, interpreted modulo-N, are generally not all different, so that a repetition of the same colors results. When a pattern contains only the single rotocenter, and there is no enantiomorphy, N will have a maximum possible value of k when no color repeats. When colors do repeat, N equals a fraction of k, so that in general $1 \leq N \leq k$. When

$p = 0$, all equivalent points have the same color, so that $N = 1$. When $p = 1$, all equivalent points have different colors, so that $N = k$. These two cases therefore represent the extreme values of N; for other values of p, N will have intermediate values.

The same color is repeated whenever p and k have a common factor greater than unity. If the greatest such factor is called u, the general item in the color sequence becomes

$$s + \frac{jpN}{k} = s + \frac{(jp/u)N}{k/u}.$$

The number u indicates the number of times the same sequence of colors is repeated around the k-fold rotocenter. When this center represents the only symmetry in the pattern, $N = k/u$, so that the general item becomes $s + jp/u$, and the color sequence becomes

$$\left[s, \quad s + \frac{p}{u}, \quad s + \frac{2p}{u}, \ldots, \quad s + \frac{jp}{u}, \ldots, \quad s + (k-1)\frac{p}{u} \right]_{\mathrm{mod}(k/u)}$$

This sequence is the algorism that generates all colored patterns having a single rotocenter with symmetry value k and color parameter p, $0 \leqslant p < k$. If p and k have no common factor greater than unity, this sequence becomes

$$[s, \quad s + p, \quad s + 2p, \ldots, \quad s + jp, \ldots, \quad s + (k-1)p]_{\mathrm{mod}\,k}.$$

In this case each color occurs only once, regardless of the value of p. Since there is no color repetition, all configurations are permutations of the same k colors. When k and p have no common factor greater than unity, all configurations having positive p therefore are equivalent, and we need only consider the one having $p = 1$.

If k equals an integral multiple of p, $u = p$, so that the same color sequence occurs p times:

$$[s, \quad s + 1, \quad s + 2, \ldots, \quad s + j, \ldots, \quad s + (k-1)]_{\mathrm{mod}\,(k/p)}$$

Figure 81 illustrates these sequences for some particular combinations of values of k and p; we have arbitrarily set $s = 0$. Note that, since in Fig. 81c the parameters k and p have no common factor greater than unity, the case $p = 3$ is entirely equivalent to the case $p = 1$ (Fig. 81a); the permutation $1 \rightarrow 3 \rightarrow 4 \rightarrow 2 \rightarrow 1$ transforms Fig. 81a into Fig. 81c.

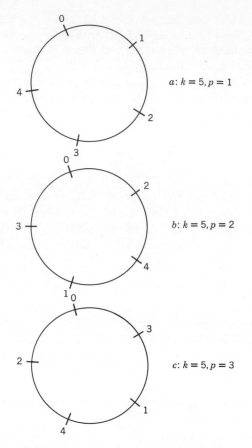

Fig. 81. Color sequence for some representative combinations of k and p. (Note that the sequences for $p = 2$ and $p = 3$ are identical but opposite in direction.)

We have noted previously (Section 12.C) that reversal of a color sequence is a particular case of permutation of colors and hence produces a pattern equivalent to the original one. If we reverse the sequence

$$s, \quad s+\frac{pN}{k}, \quad s+\frac{2pN}{k}, \dots, \quad s+\frac{jpN}{k}, \dots, \quad s+\frac{(k-1)pN}{k},$$

we obtain (not listing the general term explicitly):

$$s+\frac{(k-1)pN}{k}, \quad s+\frac{(k-2)pN}{k}, \dots, \quad s+\frac{2pN}{k}, \quad s+\frac{pN}{k}, \quad s.$$

Make a cyclical permutation, moving s to the leading position:

$$s, \quad s+\frac{(k-1)pN}{k}, \quad s+\frac{(k-2)pN}{k}, \ldots, \quad s+\frac{2pN}{k}, \quad s+\frac{pN}{k}.$$

Remembering that all terms are interpreted modulo-N, we find

$$\frac{(k-1)pN}{k} \underset{=}{N} -\frac{pN}{k} \underset{=}{N} \frac{(k-p)N}{k}, \text{ etc.,}$$

so that the sequence becomes

$$s, \quad s+\frac{(k-p)N}{k}, \quad s+\frac{2(k-p)N}{k}, \ldots, \quad s+\frac{i(k-p)N}{k}, \ldots,$$

$$s+\frac{(k-1)(k-p)N}{k}$$

This sequence, being the sequence having parameters (k, p) read in reverse direction, therefore turns out to be a sequence having parameters (k, p') read forward where p and p' are related by the identity

$$p' \underset{=}{k} k-p.$$

This identity is stated modulo k so that, when $p=0$, p' also vanishes, and p' therefore satisfies the inequalities $0 \leqslant p' < k$ common to all proper color parameters. We have therefore the following theorem.

Theorem 27. Two colored configurations about different k-fold rotocenters whose color parameters p and p' are related by the expression

$$p' \underset{=}{k} k-p$$

correspond to the same color sequence being read in opposite directions.

It should be noted that reversal of a color sequence does *not* affect the *geometrical* configuration around a rotocenter and hence should not be confused with the creation of an enantiomorphic configuration.

13

Coexistence of color rotocenters

A. The second diophantine equation

According to Theorem 1 the coexistence of two noncongruent roto-centers in a plane implies the existence of a third rotocenter in that plane; the symmetry values of a mutually implied triplet of rotocenters are constrained by a diophantine equation (1). We shall now find that the color parameters of such rotocenters are also subject to the con-straint of a diophantine equation.

Consider (Fig. 82) a point $P_0{}^s$ in a plane in which a k-fold, an l-fold, and an m-fold rotocenter coexist. The color of this point is s, and the color parameters of the three rotocenters are p, q, and r, respectively. The k-fold rotocenter relates the point $P_0{}^s$ to an equivalent point $P_1{}^{s+pN/k}$, which in turn through the l-fold rotocenter is related to $P_2{}^{s+(p/k+q/l)N}$. The m-fold rotocenter relates this last point to P_0, which therefore must have color $s + (p/k + q/l + r/m)N$. The consistency pos-tulates require that this color be the same as s, so that

$$\left(\frac{p}{k} + \frac{q}{l} + \frac{r}{m}\right)N \overset{N}{=} 0.$$

Since $p < k$, $q < l$, and $r < m$,

$$\left(\frac{p}{k} + \frac{q}{l} + \frac{r}{m}\right) < 3.$$

Therefore

$$\frac{p}{k} + \frac{q}{l} + \frac{r}{m} = 0, \quad 1, \quad \text{or} \quad 2.$$

112

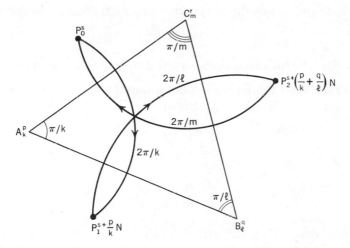

Fig. 82. Three mutually implied rotocenters, A_k^p, B_l^q, and C_m^r, whose symmetry values are denoted by subscripts, and whose color parameters are designated by superscripts.

Since p, q, and r are not negative, the sum vanishes only if all three color parameters vanish, i.e., when all congruent points have the same color. The case when the sum equals 2 can be reduced to the remaining one by the substitutions

$$p' \stackrel{k}{=} k - p, \quad q' \stackrel{l}{=} l - q, \quad r' \stackrel{m}{=} m - r;$$

we have already seen (Theorem 27) that this substitution amounts to a reversal of each color sequence, hence to a pure color transformation.

The substitution produces

$$\frac{(k-p')}{k} + \frac{(l-q')}{l} + \frac{(m-r')}{m} = 2$$

$$\therefore \frac{p'}{k} + \frac{q'}{l} + \frac{r'}{m} = 1.$$

Thus every case where the sum of p/k, q/l, and r/m equals 2 can be reduced to an equivalent one having this sum equal to unity by simply reversing the color sequence.

Theorem 28. The color parameters p, q, and r of three coexisting rotocenters in a plane either all vanish or are related by the diophantine equation

$$\frac{p}{k} + \frac{q}{l} + \frac{r}{m} = 1,$$

where k, l, and m are the respective symmetry values of these roto-centers.

B.   Solution of the second diophantine equation

Color rotational symmetry has been found to be constrained by two simultaneous diophantine equations:

$$\frac{1}{k}+\frac{1}{l}+\frac{1}{m}=1 \tag{1}$$

$$\frac{p}{k}+\frac{q}{l}+\frac{r}{m}=1 \tag{3}$$

Inspection shows that, regardless of the values of k, l, and m, there will always be a solution $p = q = r = 1$, for this solution makes the two equations identical. Theorem 28 also admits the combination $p = q = r = 0$ for all k, l, and m.

In Section 12.D, we observed that the case of color translational symmetry is fairly trivial, so that we shall not concern ourselves further with the solution $k = 1, l = \infty, m = \infty$ of (1). We shall, however, solve (3) for each of the remaining four solutions of (1). Since $p < k$, $q < l$, and $r < m$, at most one of the parameters p, q, and r in (3) may vanish at one time.

1. When

$$k = l = 2, \qquad \frac{p+q}{2}+\frac{r}{m}=1.$$

Since neither p nor q may exceed 2, and $p = q = 0$ has already been excluded, their only possible values are 0, 1; 1, 0; and 1, 1. The first two solutions are degenerate, because it does not matter which of the two twofold rotocomplexes is associated with k (and hence with p), and which with l (and hence with q). The third solution corresponds to the ubiquitous $p = q = r = 1$. Therefore the only color parameters in the $22'\infty$ system are:

p	q	r
0	0	0
0	1	$\frac{1}{2}m$
1	1	1

2. When

$$k = 2, \qquad l = 3, \qquad m = 6, \qquad \tfrac{1}{2}p + \tfrac{1}{3}q + \tfrac{1}{6}r = 1.$$

When $p = 0$, $\tfrac{1}{3}q + \tfrac{1}{6}r = 1$. Since $r < 6$, $q \neq 0$. Then when $q = 1$, $r = 4$; when $q = 2$, $r = 2$. (This solution is equivalent to the previous one by Theorem 27.)

When $p = 1$, $\tfrac{1}{3}q + \tfrac{1}{6}r = \tfrac{1}{2}$

$$\text{When } q = 0, \qquad r = 3;$$
$$\text{when } q = 1, \qquad r = 1.$$

Since p must be less than 2, and q less than 3, these solutions exhaust all possibilities for the 236 system, which are summarized as follows:

p	q	r
0	0	0
0	1	4
1	0	3
1	1	1

3. When

$$k = 2, \qquad l = m = 4, \qquad \tfrac{1}{2}p + \tfrac{1}{4}(q + r) = 1.$$

When $p = 0$, $q + r = 4$. Since neither q nor r may exceed 4, there are only the possibilities 0, 1, 3; 0, 2, 2; and 0, 3, 1. The first and the last of these are equivalent according to Theorem 27.
When $p = 1$, $q + r = 2$. There are only two distinct solutions:

$$q = 0, \qquad r = 2, \qquad \text{and} \qquad q = r = 1.$$

Since p must be less than 2, this exhausts all possibilities in the 244 system, which are summarized as follows:

p	q	r
0	0	0
0	1	3
0	2	2
1	0	2
1	1	1

4. When

$$k = l = m = 3, \qquad p + q + r = 3.$$

When $p = 0$, there is only one solution: $q = 1$, $r = 2$. (The solution 0, 2, 1 is equivalent to it according to Theorm 27.)
When $p = 1$, $q + r = 2$; there is the ubiquitous $p = q = r = 1$. The solutions $q = 0$, $r = 2$, and $q = 2$, $r = 0$ are degenerate with $p = 0$, $q = 1$, $r = 2$, because it does not matter which of the threefold roto-simplexes is associated with p, with q, and with r. Accordingly, there are only three possible combinations of color parameters in the 33'3" system:

p	q	r
0	0	0
0	1	2
1	1	1

C.   Color parameters in the 22′2″2‴ system

It is recalled that the only possibility for the existence of more than three rotosimplexes in a plane is the existence of four twofold roto-simplexes. The color parameters of these rotosimplexes are limited to values 0 and 1, hence they are binary variables. The value zero corresponds to a normal rotosimplex in Section 10.F, the value of unity to a reversing rotosimplex. Theorem 26 of the dichromatic symmetry theory can therefore be translated into the language of polychromatic symmetry.

Theorem 26a. When four rotosimplexes coexist in a plane, the sum of their color parameters must be even.

The proof of this theorem is entirely analogous to that of Theorem 26, being based on a closed circuit generated by four successive 180° rotations, and need not be given in detail here.
There are, accordingly, three possible combinations of color param-eters in the 22′2″2‴ system: 0000, 0011, and 1111.

D.   Summary

Table 9 summarizes all nonequivalent combinations of color param-eters possible in a plane.

Table 9   Symmetry values and color parameters
in the plane

Symmetry values	Rotational color parameters
$1_{\infty\infty}$	—
22_∞	000
	$00(\tfrac{1}{2}m)$
	111
236	000
	014
	103
	111
244	000
	013
	022
	102
	111
333	000
	012
	111
2222	0000
	0011
	1111

14

Color enantiomorphy

A. Introduction

In Chapter 13 we characterized the repetition of color among congruent points by the color parameters of rotocenters. In the present chapter we adjoin color enantiomorphy to congruence and examine the repetition of colors among enantiomorphic points.

When a pattern has enantiomorphy, every point not located on a mirror line has an enantiomorphic point somewhere in the pattern. Every simplex of congruent points P_0, P_1, ..., P_j, ..., has a simplex of congruent points $\hat{P}_0$, $\hat{P}_1$, ..., $\hat{P}_i$, ..., associated with it so that every point $\hat{P}_j$ is enantiomorphic to a point P_i. The colors of the simplex P_0, ..., P_j, ..., are determined by the color of a single one of its points and by the color parameters of the pattern. The colors of the simplex $\hat{P}_0$, ..., $\hat{P}_i$, ..., are determined by the color of a single one of *its* points, and by the *same* color parameters of the pattern. The color configuration of an enantiomorphic pattern is therefore completely determined by the color parameters of its rotocenters, and by the colors of a single pair of enantiomorphic points.

Suppose that a pattern contains an isochromatic pair* of enantiomorphic points P_i^s and $\hat{P}_j^s$. The same transformation that transforms P_i^s into an equivalent point $P_{i+1}^{s+\Delta s}$ will transform $\hat{P}_j^s$ into a point that is equivalent to P_j^s, and according to the consistency postulates, isochromatic to $P_{i+1}^{s+\Delta s}$.

---

*Isochromatic points are points having the same color.

Theorem 29. The existence of a single pair of isochromatic enantio-morphic points in a pattern implies that any point in the pattern equivalent to this pair has an isochromatic enantiomorphic point somewhere in the pattern.

In exhaustively generating all colored enantiomorphic configurations we shall systematically evaluate every possible isochromaticism between enantiomorphically paired points.

B.   Fundamental regions

Color symmetry is concerned with the repetition of colors among *equivalent* (congruent or enantiomorphic) points; isochromacy of *distinct* points is of no concern here. The reason for excluding the repetition of color among distinct points is that we introduced color to permit us to deal with analogous as well as identical points (cf. Section 8.4). Distinct points are not even analogous; the repetition of color among distinct points would be subject to no particular restrictions or order and would give rise to no enumerable set of configurations.

We defined *fundamental regions* in Section 3.D as members of a set of mutually equivalent regions that together cover the plane and within which no two points may be equivalent. Since no pair of points within the same fundamental region can be equivalent, we shall never be concerned about color relations between points belonging to the same fundamental regions; therefore we shall assign a common color to all points in the same fundamental region. When esthetic considerations demand more than a single color per fundamental region, the word *color* as used here may be replaced by *color combination*, with the provision that each color transformation be carried out consistently. It is essential only that a single symbol be used to denote the color(s) of any one fundamental region.

In the absence of enantiomorphy, the area of a fundamental region is uniquely determined, but its outline is not. In Fig. 83, M. C. Escher has chosen a dwarf as a fundamental region; all dwarfs are equivalent, together cover the plane, and each is in itself not symmetrical. We discussed in Section 7.I, the fact that mirror lines necessarily constitute boundaries of fundamental regions, so that when all rotocenters lie on mirror lines, each mesh constitutes a fundamental region. When k-fold rotocenters are enantiomorphically paired and do not lie on mirror lines, the mirror lines constitute boundaries between fundamental regions. Each k-fold rotocenter is then surrounded by a poly-

Fig. 83. Dwarfs, by M. C. Escher (reproduced with permission of the artist and the Escher Foundation).

gon formed by k mirrors; each such polygon can be arbitrarily divided into k congruent fundamental regions that meet at the k-fold rotocenter.

In illustrating the color-symmetrical configurations we use a single color per fundamental region. When all rotocenters are located on mirror lines, each mesh constitutes a fundamental region, so that we use a single color for each mesh. Any such illustration in which adjacent meshes are isochromatic will also serve as an illustration for the corresponding cases without enantiomorphy, for in those cases any pair of adjacent meshes may constitute a fundamental region (cf. Section 7.I).

In assigning a single color to a fundamental region we do not provide the contrasts or outlines needed to designate geometrical motifs. Although in the chapters on dichromatic symmetry (Chapters 8, 9, and 10) we did combine geometrical and color symmetry in the same illustration, we shall henceforth illustrate color symmetry only, leaving the reader to superimpose his own geometrical motifs (cf. Chapters 1 to 7). Such superposition could be achieved by carving appropriate motifs into masks, superimposing the mask onto the illustrations of color symmetry.

15

Enumeration of colored configurations having no finite rotational symmetry

A.   No congruence

We begin here with the exhaustive enumeration of all colored configurations. We shall follow the geometrical configurations listed in Table 3 systematically, evaluating all possible color combinations for each of these configurations.

The first configuration of Table 3 is the one labeled 1; here the fundamental region encompasses the entire plane, which in our terms is necessarily monochromatic.

In configuration $1m$, a mirror line divides the plane into two enantiomorphic halves, each being a fundamental region. We shall call the color of one of these 0, of the other s. When s is not equal to zero, the plane has two colors, hence $N = 2$. This means that s must be interpreted modulo-2, and since it does not equal zero, it must be equal to unity. When s does equal zero, the pattern is monochromatic, hence $N = 1$. There are therefore two colored configurations for $1m$: ($N = 2$, $s = 1$), and ($N = 1$, $s = 0$).

B.   Translational symmetry in a single direction

First comes the configuration ∞, which was discussed in Section 12.D; its color sequence is

$$[\ldots, s + \Delta s, \ldots, s + j\Delta s, \ldots]_N$$

(the subscript N indicates that all numbers in the sequence are interpreted modulo-N).

122

When enantiomorphy is adjoined to this symmetry, we must assign colors to one enantiomorphic pair of points in order to define all other colors (cf. Section 14.A). We shall set the color of one of these equal to zero, the other to s.

The configuration $\infty mm'$ is generated by two parallel mirrors, m and m'; it consists of alternating equally spaced parallel mirrors. The fundamental regions here consist of the zones between the mirrors. The translational symmetry affects the enantiomorphic pairs analogously, so that the color sequence becomes

$$[\ldots |0|s|\Delta s|s+\Delta s|2\Delta s| \ldots |s + (j-1)\Delta s|j\Delta s|s+j\Delta s| \ldots]_N$$

(vertical bars indicate mirrors). It is seen that half of the fundamental regions have colors that are functions of s, whereas the other half (their enantiomorphs) are not. Isochromatic pairs of enantiomorphic regions occur only if s has appropriate values, i.e., if it assumes the form $i\Delta s$, where i is any integer. In that case every fundamental region has an isochromatic enantiomorph (Theorem 29). When there are no isochromatic enantiomorphs, half of the fundamental regions have colors $0, \Delta s, \ldots, j\Delta s, \ldots$; their enantiomorphs have colors $s, s+\Delta s, \ldots, s \mid j\Delta s, \ldots$. Since these are all the colors occurring in the pattern, we may set Δs equal to either zero or 2, so that one half of the fundamental regions have even colors. This leaves only the odd colors for the regions enantiomorphic to them, so that s must be odd. When no enantiomorphic regions are isochromatic, all configurations having odd values of s are equivalent, so that we may set $s = 1$. There are therefore two possible colored configurations: one having $\Delta s = 0$, $s = 1$, hence $N = 2$, the other having $\Delta s = 2$, $s = 1$, and N arbitrary. In the former all congruent points are isochromatic, and enantiomorphic points have different colors.

When all enantiomorphic points are isochromatically paired, we may set $\Delta s = 1$; in that case, there are N different configurations, corresponding to all integral values of s, $0 \leqslant s < N$.

Exactly the same considerations apply to the configurations ∞m and ∞g, which correspond respectively to a single mirror line and a single glide line; these reflection lines bisect the plane into enantiomorphic halves. Each half consists of an infinite number of parallel, congruent fundamental domains, separated from each other by arbitrary, though mutually congruent boundaries. Their color sequences are, respectively:

...	0	Δs	$2\Delta s$	...	$j\Delta s$	... $\big\{$ Mirror
...	s	$s+\Delta s$	$s+2\Delta s$	...	$s+j\Delta s$	... $\big\{$ line

...0	Δs		$2\Delta s$ ...	$(j-1)\Delta s$	$j\Delta s$	... $\big\{$ Glide
	s	$s+\Delta s$	...	$Sr(j-1)\,\Delta s$	$s+j\Delta s$ ...	$\big\{$ line

Again, enantiomorphic regions are isochromatically paired if $s = i\Delta s$; we can then set $\Delta s = 1$, and let s assume all integral values from 0 up to N.

C.   Translational symmetry in two nonparallel directions

The configuration $1\infty\infty'$ was discussed in Section 12.D and offered no particular problems or challenge. The color sequences for $1\infty\infty'mm'$, $1\infty\infty'mg$ and $1\infty\infty'gg'$ are shown schematically in Fig. 84.

When no enantiomorphic regions are isochromatic, Δs and Δt may both be either zero or 2; s will then be equal to unity, in order to divide the enantiomorphs into even and odd-colored "camps." When both Δs and Δt are zero, $N = 2$; when either or both equal 2, the number of colors may be chosen arbitrarily.

D.   Summary

In adjoining enantiomorphy to translational symmetry we distinguish between the case when each color appears at least once on both sides of the reflection lines and the case when enantiomorphs are necessarily colored differently. In the former case there are numerous ways in which color can repeat among enantiomorphs. In the latter case half of the fundamental regions (all mutually congruent) are given the even colors, their enantiomorphs the odd ones; the minimum number of possible colors is then $N = 2$. In this latter case, for each of the four possible combinations of values of Δs and Δt (namely 00, 02, 22, and 20), there is only one independent colored configuration for each arbitrarily chosen value of N; all others represent color permutations of this one, and hence are equivalent to it.

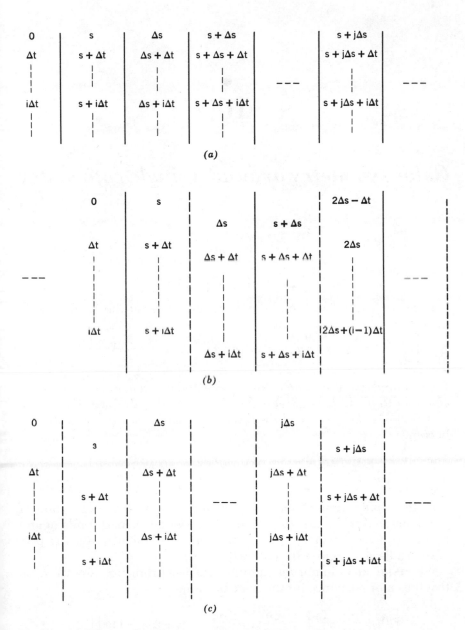

Fig. 84. (a) Colors in the 1∞∞′*mm*′ configuration. (b) Colors in the 1∞∞′*mg* configuration. (c) Colors in the 1∞∞′*gg*′ configuration.

16

Color symmetry around a single rotocenter

A. Color sequence

In Section 12.E, we found the color sequence for the configuration k:

$$\left[s,\quad s+\frac{pN}{k},\quad s+\frac{2pN}{k},\dots,\quad s+\frac{jpN}{k},\dots,\quad s+\frac{(k-1)pN}{k}\right]_N$$

It was found that the same color is repeated when u, the largest factor common to p and k, is greater than unity. In the absence of enantiomorphy, $N=k/u$, so that the color sequence for configuration k becomes

$$\left[s,\quad s+\frac{p}{u},\quad s+\frac{2p}{u},\dots,\quad s+\frac{jp}{u},\dots,\quad s+\frac{(k-1)p}{u}\right]_{\mathrm{mod}\,k/u}$$

In the configuration km, the rotocenter lies at the intersection of k mirrors and is the meeting point of $2k$ wedge-shaped fundamental regions. Adjacent regions are enantiomorphs. We shall assign to a pair of such adjacent regions the colors 0 and s.

When no enantiomorphic points are isochromatic, $N=2k/u$, so that the color sequence for this case becomes

$$\left[0\big|s\Big|\frac{2p}{u}\Big|s+\frac{2p}{u}\Big|\dots\frac{2jp}{u}\Big|s+\frac{2jp}{u}\Big|\dots\Big|s+2(k-1)\frac{p}{u}\Big|\right]_{2k/u}.$$

The colors independent of s are then all even; the remaining colors must be odd, and since all odd values of s lead to equivalent configura-

126

tions, we can set $s = 1$. Accordingly, the sequence becomes

$$\left[|0|1\left|\frac{2p}{u}\right|\frac{2p}{u}+1\right|\cdots\left|\frac{2jp}{u}\right|\frac{2jp}{u}+1\right|\cdots\left|\frac{2(k-1)p}{u}\right|1+\frac{2(k-1)p}{u}\right]_{2k/u}.$$

On the other hand, if s equals an integral multiple of pN/k, say $s = ipN/k$, enantiomorphs are isochromatically paired, so that $N = k/u$, and the color sequence becomes

$$\left[\left|0\left|\frac{ip}{u}\right|\frac{p}{u}\right|(i+1)\frac{p}{u}\right|\cdots\left|\frac{jp}{u}\right|\frac{(i+j)p}{u}\right|\cdots\left|\frac{2(k-1)p}{u}\right|\frac{(2k-2+i)p}{u}\right]_{k/u}.$$

For every combination of k and p, the integer i can range from 0 to N, giving rise to N different configurations.

B.　Illustrations

The configurations following km in Table 3 have no arbitrary parameters: they form a closed, enumerable system. Through km there is always a free parameter, whose value may be chosen arbitrarily, so that the possibilities are unlimited. In the case of translational symmetry this free parameter is the number of colors N, which is not constrained at all. For the configurations km the number of colors N is constrained by the value of k, but this number may have any positive integral value. For this reason we can only choose certain representative interesting examples to illustrate the color configurations. Figure 85 shows the colors in terms of s when $k = 5$ and $p = 2$, hence $u = 1$.

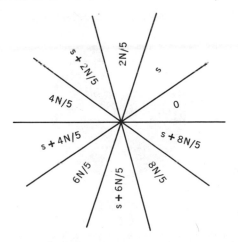

Fig. 85.　Colors around a fivefold rotocenter on mirror lines; $p = 2$.

[We showed in Theorem 27 that the case $p = 3$ amounts to a reversal of the color sequence for $p = 2$, and similarly that $p = 4$ and $p = 1$ give equivalent results. Therefore $p = 2$ provides an interesting representative case.] Since $k = 5$, the maximum number of colors is $N = 10$; this occurs when s is odd. All odd values of s lead to equivalent configurations. Five colors result when enantiomorphs are pairwise isochromatic: $N = 5$. There are several possible pairings; we should consider all values of s, $0 \leqslant s < 5$. However, inspection of Fig. 85 shows that the cases $s = 0$ and $s = 2N/5$ would both give rise to configurations in which adjacent fundamental regions are isochromatic, hence are equivalent. Similarly, the cases $s = 4N/5$ and $s = 8N/5$ are equivalent, since s is symmetrically located between $4N/5$ and $8N/5$ in

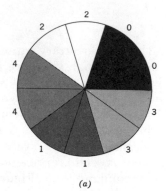

(a)

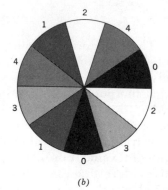

(b)

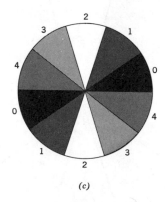

(c)

Fig. 86. Colored configurations having $k = 5$, $p = 2$, $N = 5$. (a) $s = 0, N = 5$. (b) $s = 4$, $N = 5$. (c) $s = 8, N = 5$.

Fig. 85. There are therefore three five-color configurations for the parameters $k = 5$, $p = 2$, namely, those having $s \geq 0$, $s \leq 4N/5 = 4$, and $s \geq 6N/5 \geq 1$. They are illustrated in Fig. 86.

Figure 86b is particularly interesting from the point of view of the consistency postulates. Take any one of the five mirrors, and locate two identically colored fundamental regions on one side of this mirror. When reflected in the mirror, these regions produce mirror images that also are isochromatic, in accordance with the consistency postulates. The same result is obtained for each of the five mirrors, and for every pair of isochromatic fundamental regions.

17

Enumeration of colored configurations in the 22′∞ system

A. Introduction

We now move on to the systems that have a finite, closed number of configurations. Each of these configurations will be separately generated and illustrated, in the sequence dictated by Table 3. The permitted combinations of color parameters are listed in Table 9.

In the absence of enantiomorphy the outline of the fundamental region may be arbitrarily chosen (Fig. 83). We use as a fundamental region two adjacent meshes. We shall anyhow generate all configurations in which all rotocenters lie on mirror lines; any of these in which pairs of adjacent meshes are isochromatic will also serve as illustrations of color symmetry without enantiomorphy (cf. Section 14.B.). Therefore the problem of color symmetry without enantiomorphy need not concern us explicitly.

B. The colored 22′∞ configurations

First we consider the case where all twofold rotocenters lie on mirror lines. The net is drawn in Fig. 87: the fundamental regions are bounded by the mirror lines. Colors are expressed in terms of the color parameters p and q; the colors of an adjacent enantiomorphic pair of fundamental regions are denoted by 0 and s, respectively. The colors of all other fundamental regions are now completely determined in terms of s, p, and q, as indicated in Fig. 87. For each of the

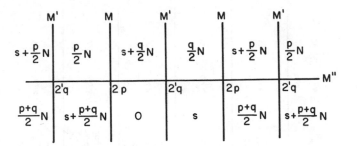

Fig. 87. Color parameters for the 22′∞ configuration.

combinations of p and q listed in Table 9 we shall consider all possible values of s.

When $p = q = 0$, the only colors are 0 and s. When $s = 0$, the pattern is monochromatic ($N = 1$). When $s \neq 0$, $N = 2$; the only possible value of s modulo-N is then unity. The resulting configuration is the dichromatic one labeled $\underline{22}'\infty$. It, and all subsequently generated configurations are illustrated in Fig. 88; this figure constitutes an exhaustive tabulation of all colored configurations in the $22'\infty$, 236, 244′, 33′3″, and 22′2″2‴ systems. Each illustration is labeled according to the entry in Table 3 to which it belongs, followed by the respective color parameters of the listed rotocenters, the number of colors (the value of N), and the value of s; these parameters uniquely and sufficiently describe the color and geometric symmetry of any pattern.

When $p = 0$, $q = 1$, the colors are 0, s, $\frac{1}{2}N$, $s + \frac{1}{2}N$. When all these expressions have different values, the maximum number of colors, $N = 4$, occurs. Since in this case $\frac{1}{2}N = 2$, s is limited to the values 1 and 3; these two cases are equivalent, so that we have a single four-colored configuration: $\underline{22}'\infty$, 01, $N = 4$, $s = 1$. When the four expressions are equal pairwise, $N = 2$; s may then equal 0 or 1; two two-colored configurations result: $\underline{22}'\infty$, 01, $N = 2$, $s = 0$, and $\underline{22}'\infty$, 01, $N = 2$, $s = 1$.

When $p = q = 1$, the colors are likewise 0, s, $\frac{1}{2}N$, $s + \frac{1}{2}N$, although differently distributed over the fundamental regions. The analysis of the previous case applies equally, so that here also a single four-colored and two two-colored configurations result. All of these configurations are represented in Fig. 88.

C. The colored $2\hat{2}\,\infty$ configurations

This configuration is shown in terms of its color parameters in Fig. 89. Mirrors perpendicularly bisect the sections of a glide line that join the twofold rotocenters. The glide line is not shown explicitly,

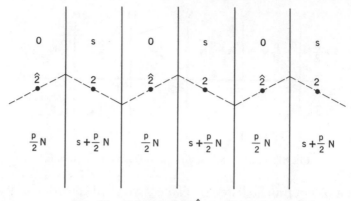

Fig. 89.   Colors in the $2\hat{2}\infty$ configuration.

for the fundamental regions are bounded by mirror lines, but not necessarily by the glide line. The boundary shown is arbitrarily chosen for this illustration.

The color parameters of the rotosimplexes may both be equal to zero or to one: here both color parameters are binary variables. A color parameter that equals zero relates points of equal color, whereas a unit parameter relates differently colored points. Since the two roto-complexes are enantiomorphically paired, consistency requires that both have either zero or unit color parameters. Therefore the color parameters of the two enantiomorphically paired rotocomplexes can both be set equal to p, which may equal either zero or unity. Accordingly, the colors are expressed in Fig. 89 in terms of N, s, and p; these three parameters uniquely describe the color configurations of any 22∞ pattern.

When $p = 0$, the only colors are 0 and s; there is, accordingly, a monochromatic configuration $N = 1$, $s = 0$, and a dichromatic one $N = 2$, $s = 1$. These are illustrated in Fig. 88 which is a 16-page color insert following page 175.

When $p = 1$, the colors are 0, s, $\frac{1}{2}N$, $s + \frac{1}{2}N$. As in the case of the $\underline{22'}\infty$ configurations, there is a single four-colored configuration: $N = 4$, $s = 1$, and there are two two-colored configurations: $N = 2$, $s = 0$, and $N = 2$, $s = 1$. All of these are illustrated in Fig. 88.

D.   Summary of colored 22∞ configurations

Table 10 lists exhaustively the colored configurations possible in the 22∞ system. They are illustrated in Fig. 88.

Table 10   Colored 22∞ configurations

| First rotocenter | | Second rotocenter | | | |
Symmetry value	Color parameter	Symmetry value	Color parameter	N	s
2	$\begin{cases} 0 \\ \\ \\ 0 \\ \\ \\ 1 \end{cases}$	$2'$	$\begin{cases} 0 \\ \\ 1 \\ \\ \\ 1 \end{cases}$	$\begin{cases} 2 \\ 1 \end{cases}$ $\begin{cases} 4 \\ 2 \end{cases}$ $\begin{cases} 4 \\ 2 \end{cases}$	1 0 1 $\begin{cases} 0 \\ 1 \end{cases}$ 1 $\begin{cases} 0 \\ 1 \end{cases}$
2	$\begin{cases} 0 \\ \\ 1 \end{cases}$	$\hat{2}$	$\begin{cases} 0 \\ \\ 1 \end{cases}$	$\begin{cases} 2 \\ 1 \end{cases}$ $\begin{cases} 4 \\ 2 \end{cases}$	1 0 1 $\begin{cases} 0 \\ 1 \end{cases}$

18

Enumeration of colored configurations in the 236 system

A.   Colors in the 236 net

Since no rotocenters can be enantiomorphically paired, these configurations either have no enantiomorphy or have all rotocenters located on mirror lines. Accordingly, all meshes have a single color. Table 9 lists four permitted combinations of color parameters for the 236 system. Figure 90 shows the colors in the 236 net as functions of N, p, q, and s. When $p = q = 0$, the only colors are 0 and s; there is a monochromatic configuration $N = 1$, $s = 0$, and a dichromatic one: $N = 2$, $s = 1$.

When $p = 0$, $q = 1$, the colors are: $0, s, \frac{1}{3}N, s + \frac{1}{3}N, \frac{2}{3}N, s + \frac{2}{3}N$. When all these are different, $N = 6$; s must then be odd, and since all odd values of s give rise to equivalent configurations, there is but a single six-colored configuration: $N = 6$, $s = 1$. When the six colors are equal pairwise, $N = 3$, and there are three different configurations, having respectively $s = 0, 1$, and 2.

When $p = 1$, $q = 0$, the colors are $0, s, \frac{1}{2}N, s + \frac{1}{2}N$. As before, there is a single four-colored configuration: $N = 4$, $s = 1$. There are two dichromatic configurations: $N = 2$, $s = 0$, and $N = 2$, $s = 1$.

When $p = q = 1$, there are twelve colors:

$$0, \quad s, \quad \tfrac{1}{6}N, \quad s + \tfrac{1}{6}N, \quad \tfrac{1}{3}N, \quad s + \tfrac{1}{3}N,$$

$$\tfrac{1}{2}N, \quad s + \tfrac{1}{2}N, \quad \tfrac{2}{3}N, \quad s + \tfrac{2}{3}N, \quad \tfrac{5}{6}N, \quad s + \tfrac{5}{6}N.$$

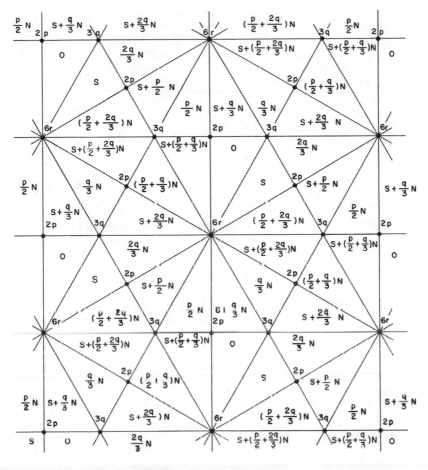

Fig. 90.   Colors in the 236 net.

When all of these are different, a twelve-colored configuration results: $N = 12$, $s = 1$. When these twelve are pairwise equal, six colored configurations arise: $N = 6$, $s = 0$; $N = 6$, $s = 1$; $N = 6$, $s = 2$; $N = 6$, $s = 3$; $N = 6$, $s = 4$; $N = 6$, $s = 5$. Since all colors must have integral values, N cannot be less than 6 for the color parameters $p = q = 1$.

B.   Summary

All colored configurations are summarized in Table 11 and illustrated in Fig. 88.

Table 11   Colored 236 configurations

Rotocenters

Symm. value	Color param.	Symm. value	Color param.	Symm. value	Color param.	N	s
2	0	3	0	6	0	$\begin{cases} 2 \\ 1 \end{cases}$	$\begin{matrix} 1 \\ 0 \end{matrix}$
	0		1		4	$\begin{cases} 6 \\ 3 \end{cases}$	$\begin{matrix} 1 \\ \begin{cases} 0 \\ 1 \\ 2 \end{cases} \end{matrix}$
	1		0		3	$\begin{cases} 4 \\ 2 \end{cases}$	$\begin{matrix} 1 \\ \begin{cases} 0 \\ 1 \end{cases} \end{matrix}$
	1		1		1	$\begin{cases} 12 \\ 6 \end{cases}$	$\begin{matrix} 1 \\ \begin{cases} 0 \\ 1 \\ 2 \\ 3 \\ 4 \\ 5 \end{cases} \end{matrix}$

19

Enumeration of colored configurations in the 244' system

A. Colors in the 244' net

Table 9 gives five possible combinations of color parameters for this system. The colors are given in terms of these parameters and of N and s in Fig. 91.

When $p = q = r = 0$, the colors are 0 and s; there is a monochromatic configuration $N = 1$, $s = 0$, and a dichromatic one $N = 2$, $s = 1$.

When $p = 0$, $q = 1$, the colors are 0, s, $\frac{1}{4}N$, $s + \frac{1}{4}N$, $\frac{1}{2}N$, $s + \frac{1}{2}N$, $\frac{3}{4}N$, $s + \frac{3}{4}N$. There is, when all these are different, an eight-colored configuration $N = 8$, $s = 1$, and when all eight are equal pairwise, there are four four-colored ones: $N = 4$, $s = 0$; $N = 4$, $s = 1$; $N = 4$, $s = 2$; $N = 4$, $s = 3$.

When $p = 0$, $q = 2$, the colors are $0, s, \frac{1}{2}N, s + \frac{1}{2}N$. As usual, these give rise to a four-colored configuration $N = 4$, $s = 1$, and two two-colored ones: $N = 2$, $s = 0$ and $N = 2$, $s = 1$.

When $p = 1$, $q = 0$, the colors are the same as in the preceding case, though distributed differently. Again, there are a single four-colored and two dichromatic configurations.

When $p = q = 1$, the colors are the same as in the case $p = 0$, $q = 1$. The two fourfold rotocomplexes here have the same symmetry value as well as the same color parameter, so that it is entirely immaterial which of the two is distinguished by the prime. Therefore the case $N = 4$, $s = \frac{1}{2}pN = 2$ is degenerate with $N = 4$, $s = 0$ (cf. Fig. 91). There is, accordingly, a single eight-color configuration $N = 8$, $s = 1$, and

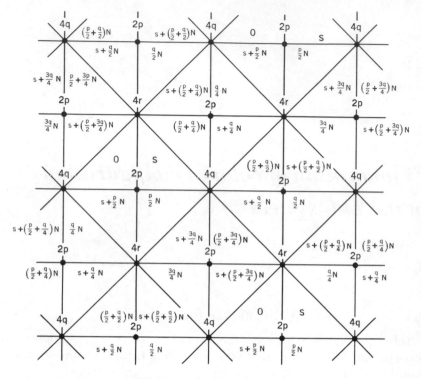

Fig. 91.   Colors in the 244′ net.

there are *three* four-colored configurations: $N = 4$, $s = 0$; $N = 4$, $s = 1$; $N = 4$, $s = 3$.

All of these configurations are shown in Fig. 88.

B.   Enantiomorphic fourfold rotosimplexes

We saw in considering the $2\hat{2}\infty$ configurations that the consistency postulates impose a restriction on the color parameters of enantiomorphic twofold rotocenters: these parameters had to be equal to each other. Similarly, we shall find a restriction on enantiomorphically coupled fourfold rotocenters. In Fig. 92 we show two enantiomorphically paired fourfold rotocenters, one having color parameter q, the other color parameter r. A point P_0 is chosen, having color 0; its mirror image is called $\hat{P}_0$, having color s. Points congruent to P_0 are called P_1, P_2, P_3, and their respective mirror images are $\hat{P}_1$, $\hat{P}_2$, $\hat{P}_3$. The colors of these points are determined by the color parameters q and r.

When all eight points related by the fourfold rotational symmetry

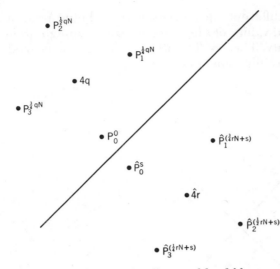

Fig. 92.   Two enantiomorphically paired fourfold rotocenters.

and the enantiomorphy have different colors, consistency is certainly
not violated. This is the case when $q = r = 1$ (the case 244, 111 in
Table 9) as long as $s \neq \frac{1}{4}jpN$. On the other hand, if $q = 0$, the four
points P_0, P_1, P_2, and P_3 have the same color, with the result that consistency is satisfied only if their enantiomorphs are isochromatic,
hence if $r = 0$. Thus the combination $2\underline{4}\hat{4}$, 102 in Table 9 is eliminated
for $2\underline{4}\hat{4}$ configurations, but $2\underline{4}\hat{4}$, 000 is allowed. The combination $2\underline{4}\hat{4}$,
022 is likewise permitted, because here the points P_0 and P_2 are iso-
chromatic as are $\hat{P}_0$ and $\hat{P}_2$, P_1 and P_3, and $\hat{P}_1$ and $\hat{P}_3$.

We shall therefore consider the configurations $2\underline{4}\hat{4}$, 000, $2\underline{4}\hat{4}$, 013;
$2\underline{4}\hat{4}$, 022; $2\underline{4}\hat{4}$, 111.

C.   Colors in the $2\underline{4}\hat{4}$ system

When $p = q = r = 0$, the colors are as indicated in Fig. 93a. The fourfold rotocenters are surrounded by squares of mirror lines that constitute boundaries of fundamental regions. Each of these squares can
be arbitrarily subdivided into four mutually congruent fundamental
regions, as long as adjacent squares are enantiomorphic. This division
is illustrated in Fig. 93; in Fig. 93a each square is monochromatic,
because all color parameters equal zero. There is a monochromatic
configuration $N = 1$, $s = 0$, and a dichromatic one $N = 2$, $s = 1$.

When $p = 0$, $q = 1$, $r = 3$, the colors are as indicated in Fig. 93b.
Eight colors can occur when all expressions in the fundamental

regions are different; this necessitates s being odd, and since, as before, all odd values of s lead to equivalent configurations we can set s equal to unity. This configuration is thus characterized as $\underline{2}4\hat{4}, 013$, $N = 8, s = 1$, as illustrated in Fig. 88.

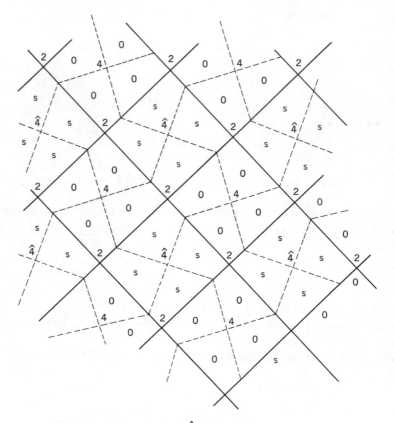

Fig. 93a. Colors in the $\underline{2}4\hat{4}$ system when $p = q = r = 0$.

When $s = \frac{1}{4}jN$, $j = 0, 1, 2, 3$, there are isochromatic enantiomorphs. However, the consistency postulates eliminate some values of j, as will now be demonstrated. Consider, in Fig. 93b, a point in a funda- mental region that has color s; its images in every mirror line have color 0. A point having color $\frac{1}{4}jN$ has mirror images having color $s + \frac{1}{4}jN$. Therefore, if points having colors s and $\frac{1}{4}jN$ are made isochro- matic, consistency demands that their mirror images be isochromatic as well. Thus $s\overset{N}{=}\tfrac{1}{4}jN$ implies $0\overset{N}{=}s+\tfrac{1}{4}jN$, and vice versa. Symbolically we say

$$s\overset{N}{=}\tfrac{1}{4}jN \iff 0\overset{N}{=}s+\tfrac{1}{4}jN$$

(the symbol $\iff$ means "implies and is implied by").

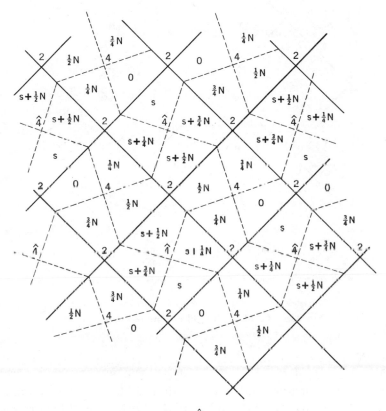

Fig. 93b.   Colors in the $\underline{2}4\hat{4}$ system: $p = 0, q = 1, r = 3$.

The two equations on both sides of the $\iff$ symbol must both be valid simultaneously; this is true if and only if $j = 0$ or 2. In these two cases $N = 4$, so that $s = 0$ or 2 also. There are therefore two four-colored configurations: $\underline{2}4\hat{4}$, 013, $N = 4$, $s = 0$, and $\underline{2}4\hat{4}$, 013, $N = 4$, $s = 2$. Both are illustrated in Fig. 88.

The colors for $\underline{2}4\hat{4}$, 022 are indicated as functions of s and N in Fig. 93c. These colors are 0, s, $\tfrac{1}{2}N$, $s+\tfrac{1}{2}N$, allowing a maximum of four

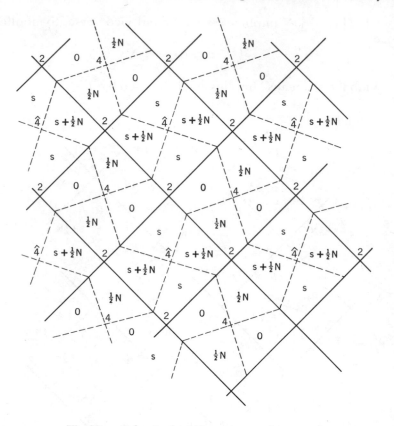

Fig. 93c.   Colors in the $\underline{2}44$ system: $p = 0$, $q = r = 2$.

colors: $N = 4$, $s = 1$. Two-color configurations occur when $N = 2$, $s = 0$ and $N = 2$, $s = 1$; all three are illustrated in Fig. 88.

Finally, we consider the configurations $\underline{2}4\hat{4}$, 111; their colors are expressed in terms of N and s in Fig. 93d. There are eight expressions: 0, s, $\frac{1}{4}N$, $s + \frac{1}{4}N$, $\frac{1}{2}N$, $s + \frac{1}{2}N$, $\frac{3}{4}N$, $s + \frac{3}{4}N$, hence a possibility for eight colors: $N = 8$, $s = 1$. Four-color configurations arise when enantiomorphs are isochromatically paired. As in the case of the $\underline{2}44$, 013 configurations (cf. Fig. 93d):

$$s \overset{N}{=} \tfrac{1}{4}jN \Longleftrightarrow 0 \overset{N}{=} s - \tfrac{1}{4}jN.$$

The two equations on both sides of the $\Longleftrightarrow$ symbol are in this case equivalent, so that all integral values of s from 0 to N are permitted.

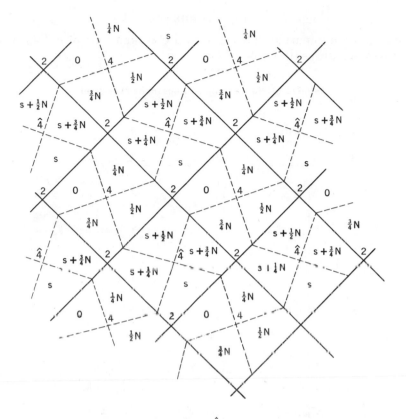

Fig. 93d.   Colors in the 244̂ system: $p = q = r - 1$.

Inspection of Fig. 93d shows that reflection in one mirror has a different result from reflection in a mirror perpendicular to the first one: a point having color s has a mirror image of color 0 in one mirror, an image of color $\frac{1}{2}N$ in the other. Nothing in the consistency postulates is at variance with this phenomenon. The configuration 244̂, $N = 4$, $s = 2$ is quite interesting in that one set of mirrors transforms colors 1 and 3 into each other, leaving 0 and 2 unaffected, whereas the mirrors perpendicular to these transform colors 0 and 2 into each other, leaving 1 and 3 unaffected. This phenomenon also is perfectly in accord with the consistency postulates, which merely require isochromatic points to remain isochromatic, but it does show that transformations of colors differ from those of geometrical coordinates. All 244̂, 111 configurations, namely $N = 8$, $s = 1$, $N = 4$, $s = 0$, $N = 4$, $s = 1$, $N = 4$, $s = 2$ and $N = 4$, $s = 3$ are illustrated in Fig. 88.

D. Summary

Table 12 summarized all permitted combinations of colors in the 244 system; all are illustrated in Fig. 88.

Table 12   Colored configurations in the 244 system

		Rotocenters					
Symm. value	Color param.	Symm. value	Color param.	Symm. value	Color param.	N	s
$\underline{2}$	0	4	0	4′	0	2	1
						1	0
	0		1		3	8	1
						4	0, 1, 2, 3
	0		2		2	4	1
						2	0, 1
	1		0		2	4	1
						2	0, 1
	1		1		1	8	1
						4	0, 1, 3
$\underline{2}$	0	4	0	$\hat{4}$	0	2	1
						1	0
	0		1		3	8	1
						4	0, 2
	0		2		2	4	1
						2	0, 1
	1		1		1	8	1
						4	0, 1, 2, 3

20

Enumeration of colored configurations in the $\underline{33'}\,\underline{3}''$ system

A. Colors in the $\underline{33}\ \underline{3}''$ net

Next in Table 3 are the configurations that have three threefold rotocomplexes. Only three combinations of color parameters are permitted (cf. Table 9). 000, 012, and 111. Figure 94 shows the colors in the $\underline{33'3}''$ net as functions of the color parameters, and of N and s.

When $p = q = 0$, there is monochromatic configuration $N = 1, s = 0$, and a dichromatic one, $N = 2, s = 1$. The latter is illustrated in Fig. 88.

When $p = 0, q = 1$, the colors are, in terms of N and s: $0, s, \frac{1}{3}N, s+\frac{1}{3}N$, $\frac{2}{3}N, s+\frac{2}{3}N$. When these are all different, a six-color pattern arises: $N = 6, s = 1$, illustrated in Fig. 88. Enantiomorphic poly-dichromatism produces three different three-colored configurations, $N = 3, s = 0$; $N = 3, s = 1$; and $N = 3, s = 2$. All of these are found in Fig. 88.

When $p = q = r = 1$, all rotocenters have the same symmetry value, as well as the same color parameter, so that it is entirely arbitrary which has the prime, and which the double prime. Therefore there is considerable degeneracy in these configurations: there is a single six-colored one, $N = 6, s = 1$, and a single three-color one: $N = 3, s = 0$, to which $N = 3, s = 1$, and $N = 3, s = 2$ are equivalent. These are both illustrated in Fig. 88.

B. Colored $3\hat{3}\underline{3}'$ configurations

The enantiomorphically paired rotocenters are surrounded by equilateral triangles of mirror lines that may be divided arbitrarily into

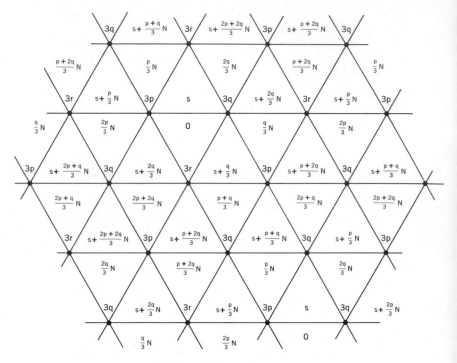

Fig. 94.   Colors in the 33'3" net.

mutually congruent fundamental regions; this has been done in Fig. 95. When $p = q = 0$, all fundamental regions within the same triangle of mirrors are isochromatic (Fig. 95a), so that there is a possible dichromatic configuration, $N = 2$, $s = 1$, and monochromatic one, $N = 1$, $s = 0$. The former is found in Fig. 88.

Figure 95b shows the colors when $p = 0$, $q = 1$. There are six colors: 0, s, $\frac{1}{3}N$, $s+\frac{1}{3}N$, $\frac{2}{3}N$, $s+\frac{2}{3}N$. When all are different, a six-colored configuration results: $N = 6$, $s = 1$, illustrated in Fig.88. When they are isochromatic pairwise, three-color configurations arise. According to the consistency postulates,

$$s \overset{N}{=} \tfrac{1}{3}jN \iff 0 \overset{N}{=} s + \tfrac{1}{3}jN.$$

These two equations can only be simultaneously valid if $j = 0$, hence $s = 0$. There is therefore only one three-color configuration: $N = 3$, The colors for the color parameters $p = q = 1$ are shown in Fig. 95c;

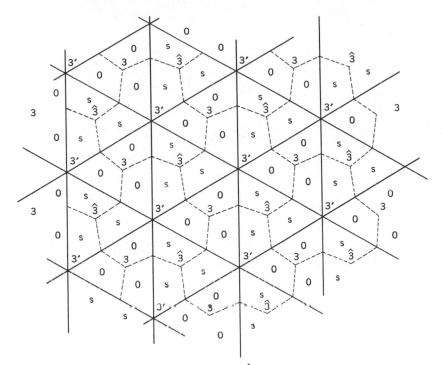

Fig. 95a. $\underline{3}'3\hat{3}, 000$

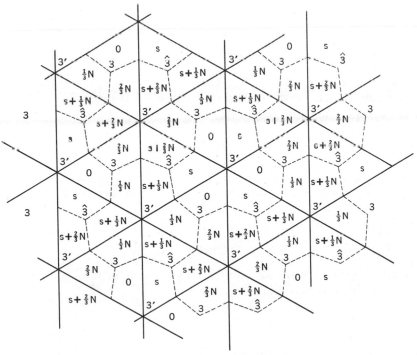

Fig. 95b. $\underline{3}'3\hat{3}, 012.$

147

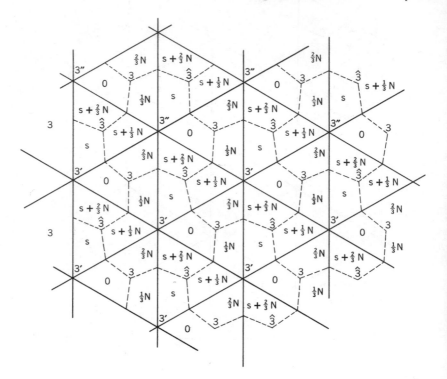

Fig. 95c.   3'33, 111.

there is, as before, a six-color configuration $N = 6$, and when enantio-morphs are isochromatically paired, there are three-color configurations. Consistency here requires

$$s \overset{N}{=} \tfrac{1}{3}jN \iff 0 \overset{N}{=} s - \tfrac{1}{3}jN,$$

which is tautological, i.e., it holds for all three possible values of s. There are, accordingly, three three-color configurations: $N = 3, s = 0$; $N = 3, s = 1$; $N = 3, s = 2$.

C.   Summary

Table 13 enumerates all color configurations in the 33'3" system; these are illustrated in Fig. 88.

Table 13 Colored configurations in the 33′3″ system

		Rotocenters					
Symm. value	Color param.	Symm. value	Color param.	Symm. value	Color param.	N	s
$\underline{3}$	$\begin{cases}0 \\[6pt] 0 \\[6pt] 1\end{cases}$	$\underline{3}'$	$\begin{cases}0 \\[6pt] 1 \\[6pt] 1\end{cases}$	$\underline{3}''$	$\begin{cases}0 \\[6pt] 2 \\[6pt] 1\end{cases}$	$\begin{cases}2\\1\end{cases}$ $\begin{cases}6\\3\end{cases}$ $\begin{cases}6\\3\end{cases}$	$\begin{matrix}1\\0\end{matrix}$ $\begin{matrix}1\\\{^0_1{}_2}\end{matrix}$ $\begin{matrix}1\\0\end{matrix}$
$\underline{3}'$	$\begin{cases}0 \\[6pt] 0 \\[6pt] 1\end{cases}$	3	$\begin{cases}0 \\[6pt] 1 \\[6pt] 1\end{cases}$	$\hat{3}$	$\begin{cases}0 \\[6pt] 2 \\[6pt] 1\end{cases}$	$\begin{cases}2\\1\end{cases}$ $\begin{cases}6\\3\end{cases}$ $\begin{cases}6\\3\end{cases}$ $\begin{cases}6\\3\end{cases}$	$\begin{matrix}1\\0\end{matrix}$ $\begin{matrix}1\\0\end{matrix}$ $\begin{matrix}0\\0\end{matrix}$ $\{^0_1{}_2$

149

21

Enumeration of colored configurations in the $22'2''2'''$ system

A. Introduction

Table 3 lists four configurations having enantiomorphy in the $22'2''2'''$ system: $\underline{22}'\underline{2}''2'''$, $2\hat{2}\underline{2}'\hat{\underline{2}}''$, $2\hat{2}2'2'g/g'$, and $2\hat{2}2'\hat{2}'m/g$. As before, the nonenantiomorphic configuration $22'2''2'''$ does not need to be considered separately from the point of view of color symmetry, because any enantiomorphic configuration in which adjacent fundamental regions are isochromatic will also serve for the $22'2''2'''$ configuration. This case differs in one respect from the preceding ones, though. Whereas the rotocenters in the other system have definite angular positions with respect to each other, the $22'2''2'''$ net is arbitrary in the absence of enantiomorphy. We have seen that enantiomorphy limits the $22'2''2'''$ net to rectangular or rhombic configurations. However, since the shape of a mesh does not affect color symmetry, there is still no reason to consider the $22'2''2'''$ configuration separately.

Table 9 gives the following combinations of color parameters for this system: all equal to zero, all equal to unity, or two equal to zero, the remaining ones equal to unity.

In the $\underline{22}'\underline{2}''2'''$ configuration each rectangular mesh has four distinct rotocenters at its vertices (cf. Fig. 52). When two of the color parameters equal zero, while the other two equal unity, there are two possibilities: diagonally opposite rotocenters may have the same or different color parameters. Thus there are four combinations: 0000, 0011, 0110, 1111.

In Section 17.C, it was noted that consistency requires enantio-morphically paired twofold rotocenters to have the same color parame-ter. In the $2\hat{2}2'2''$ configuration there are, therefore, these possibilities: 0000, 0011, 1100, 1111. In both $2\hat{2}2'\hat{2}'$ configurations there are the combinations 0000, 0011, 1111.

B.   Colors in the $\underline{22}'2''2'''$ configurations

Figure 96a shows the colors in the $\underline{22}'\underline{2}''\underline{2}'''$ net for the color parame-ters 0000. When $s \neq 0$, $N = 2$, hence $s = 1$. This produces a dichro-matic configuration, illustrated in Fig. 88. When $s = 0, N = 1$.

Figure 96b shows the colors for the parameters 0011. There are four colors, 0, s, $\frac{1}{2}N$, and $s + \frac{1}{2}N$; therefore, there is a single four-colored configuration, $N = 4$, $s = 1$. When they are equal pairwise, there are two dichromatic configurations: $N = 2$, $s = 0$, and $N = 2$, $s = 1$; all are illustrated in Fig. 88.

Figure 96c has the colors for the combination 0110. Again, there is a

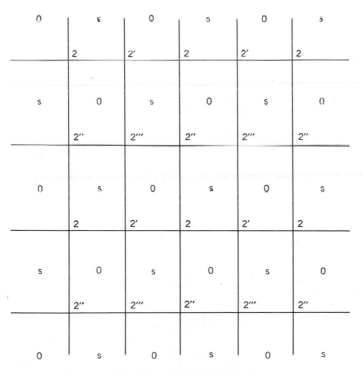

Fig. 96a. $\underline{22}'\underline{2}''\underline{2}'''$, 0000.

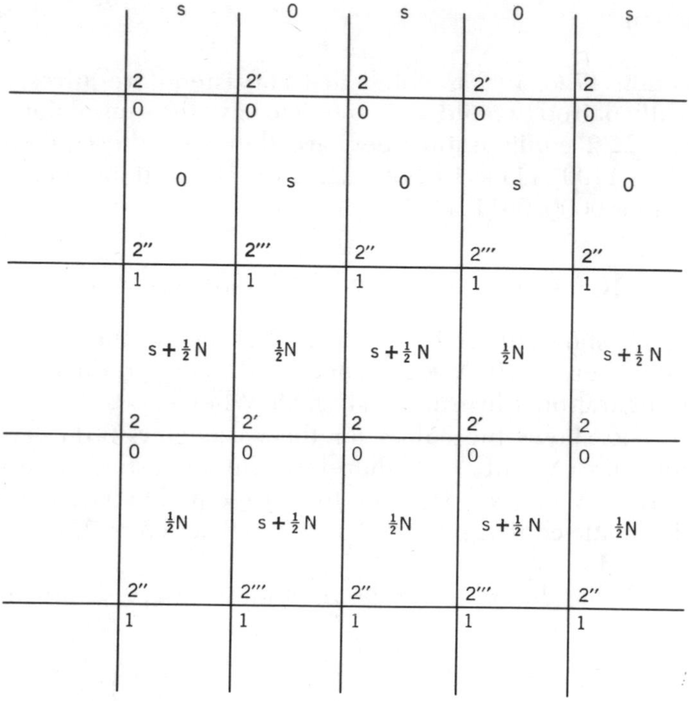

s	0	s	0	s
2	2′	2	2′	2
0	0	0	0	0
0	s	0	s	0
2″	2‴	2″	2‴	2″
1	1	1	1	1
$s+\frac{1}{2}N$	$\frac{1}{2}N$	$s+\frac{1}{2}N$	$\frac{1}{2}N$	$s+\frac{1}{2}N$
2	2′	2	2′	2
0	0	0	0	0
$\frac{1}{2}N$	$s+\frac{1}{2}N$	$\frac{1}{2}N$	$s+\frac{1}{2}N$	$\frac{1}{2}N$
2″	2‴	2″	2‴	2″
1	1	1	1	1

Fig. 96b. $\underline{2}\underline{2}'2''2'''$, 0011.

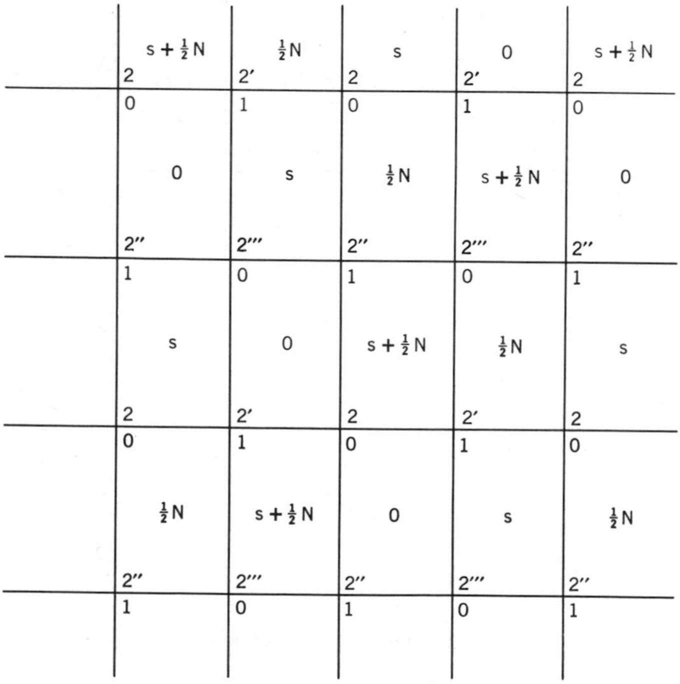

$s+\frac{1}{2}N$	$\frac{1}{2}N$	s	0	$s+\frac{1}{2}N$
2	2′	2	2′	2
0	1	0	1	0
0	s	$\frac{1}{2}N$	$s+\frac{1}{2}N$	0
2″	2‴	2″	2‴	2″
1	0	1	0	1
s	0	$s+\frac{1}{2}N$	$\frac{1}{2}N$	s
2	2′	2	2′	2
0	1	0	1	0
$\frac{1}{2}N$	$s+\frac{1}{2}N$	0	s	$\frac{1}{2}N$
2″	2‴	2″	2‴	2″
1	0	1	0	1

Fig. 96c. $\underline{2}\underline{2}'2''2'''$, 0110.

four-color configuration, $N = 4$, $s = 1$. When $s = 0$, there is a dichromatic configuration: $N = 2$, $s = 0$. Inspection of Fig. 96c shows that s is, both vertically and horizontally, surrounded by color 0 on one side, by color $\frac{1}{2}N$ on the other. The case $s = \frac{1}{2}N = 1$ is therefore equivalent to $s = 0$, so that there is but a single dichromatic configuration. Both the four-colored and the dichromatic configurations are displayed in Fig. 88.

Figure 96d shows the combination 1111; as before, there is a four-color configuration $N = 4$, $s = 1$. It is observed that in Fig. 96d the case $s = 0$ would produce horizontal colored stripes and the case $s = \frac{1}{2}N = 1$ would produce vertical ones. Since there is no particular distinction in symmetry between these two directions, these cases are equivalent, so that there is but a single dichromatic configuration: $N - 2$, $s - 0$. Both are illustrated in Fig. 88.

C. Colors in the $2\hat{2}2'2''$ configuration

Here the mirrors divide the plane into rectangles, each of which consists of two fundamental regions. At the center of each rectangle is a twofold rotocenter, located on an otherwise arbitrary boundary

$s+\frac{1}{2}N$	$\frac{1}{2}N$	$s+\frac{1}{2}N$	$\frac{1}{2}N$	
2	2'	2	2'	2
1	1	1	1	1
0	s	0	s	0
2"	2'''	2"	2'''	2"
1	1	1	1	1
$s+\frac{1}{2}N$	$\frac{1}{2}N$	$s+\frac{1}{2}N$	$\frac{1}{2}N$	$s+\frac{1}{2}N$
2	2'	2	2'	2
1	1	1	1	1
0	s	0	s	0
2"	2'''	2"	2'''	2"
1	1	1	1	1

Fig. 96d. 22'2"2''', 1111.

between two fundamental regions (cf. Fig. 97). The color-parameter combination 0000 is shown in Fig. 97a; as usual there are two configurations: $N = 2$, $s = 1$ and $N = 1$, $s = 0$, the former being illustrated in Fig. 88. The combination 0011 is shown in Fig. 97b: there are the configurations $N = 4$, $s = 1$ and $N = 2$, $s = 0$ (Fig. 88). The configuration $N = 2$, $s = 1$ is equivalent to $N = 2$, $s = 0$, as can be seen from Fig. 97b: when $s = 0$ there are isochromatic vertical stripes, whereas when $s = 1$, these stripes are horizontal.

The combination 1100 is shown in Fig. 97c, and, as before, there are the configurations $N = 4$, $s = 1$; $N = 2$, $s = 0$; and $N = 2$, $s = 1$, all illustrated in Fig. 88. Finally, Fig. 97d contains the colors for the combination 1111, which again lead to configurations $N = 4$, $s = 1$; $N = 2$, $s = 0$ and $N = 2$, $s = 1$, as illustrated in Fig. 88.

D.   Colors in the $2\hat{2}2'\hat{2}'g/g$ configuration

This configuration has no mirror lines, so that the boundaries of the fundamental regions can be drawn arbitrarily, as shown in Fig. 98. The combination 0000 (Fig. 98a) yields the dichromatic configuration $N = 2$, $s = 1$ (Fig. 88), and the monochromatic one $N = 1$, $s = 0$. The combination 0011 (Fig. 98b) yields $N = 4$, $s = 1$ and $N = 2$, $s = 0$.

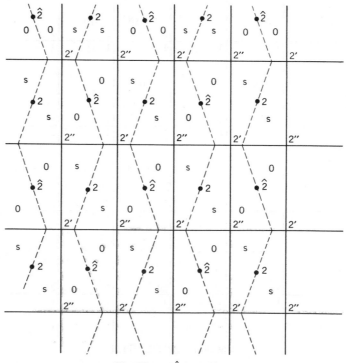

Fig. 97a. $2\hat{2}\,2'\underline{2}''$, 0000.

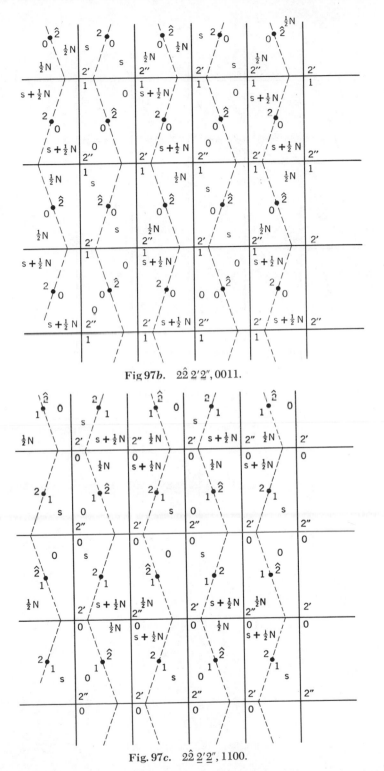

Fig 97b. $2\hat{2}\,2'2''$, 0011.

Fig. 97c. $2\hat{2}\,\underline{2}'\underline{2}''$, 1100.

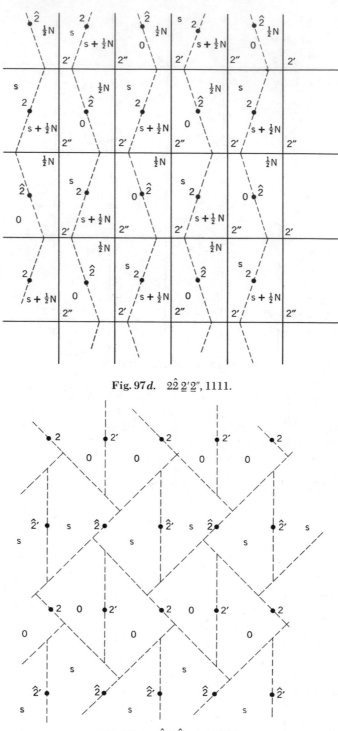

Fig. 97d. 2$\hat{2}$ $\underline{2}'\underline{2}''$, 1111.

Fig. 98a. 2$\hat{2}$ 2'$\hat{2}'$ g/g', 0000.

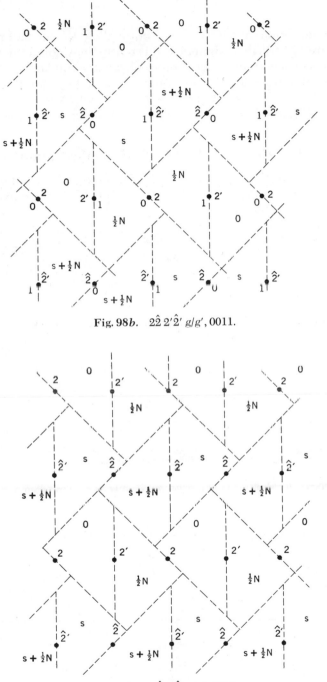

Fig. 98b. $2\hat{2}\,2'\hat{2}'\,g/g'$, 0011.

Fig. 98c. $2\hat{2}\,2'\hat{2}'\,g/g'$, 1111.

The configuration $N = 2$, $s = 1$ is equivalent to the latter, for the primes are arbitrarily assigned, and it really does not matter whether 2 or 2′ lies on a boundary between isochromatic fundamental regions. Figure 98c shows colors for the combination 1111; again there are the configurations $N = 4$, $s = 1$ and $N = 2$, $s = 0$, with $N = 2$, $s = 1$ equivalent to $N = 2$, $s = 0$. Figure 88 contains the two nonequivalent configurations.

E.   Colors in the $2\hat{2}2'\,\hat{2}'\,m/g$ configuration

The final configuration in Table 3 is $2\hat{2}2'\hat{2}'\,m/g$, whose color parameters may be 0000, 0011, or 1111. Figure 99 shows the mirror lines,

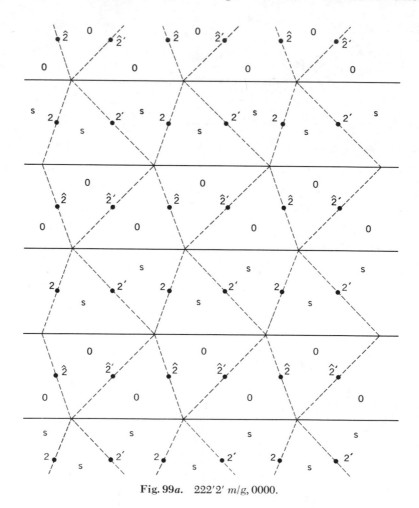

Fig. 99a.   222′2′ m/g, 0000.

all parallel to each other and bounding fundamental regions; the twofold rotocenters lie on otherwise arbitrary boundaries of fundamental regions (Fig. 99). The color-parameter combination 0000 is shown in Fig. 99a; it generates a dichromatic configuration $N = 2$, $s = 1$ (Fig. 88) as well as the monochromatic one $N = 1$, $s = 0$. The combination 0011 is illustrated in Fig. 99b, and is seen to generate the four-color configuration $N = 4$, $s = 1$, and two two-color ones $N = 2$, $s = 0$ and $N = 2$, $s = 1$ (Fig. 88). Similarly the combination 1111 (Fig. 99c) generates a four-color configuration $N = 4$, $s = 1$, and two two-color ones, $N = 2$, $s = 0$ and $N = 2$, $s = 1$.

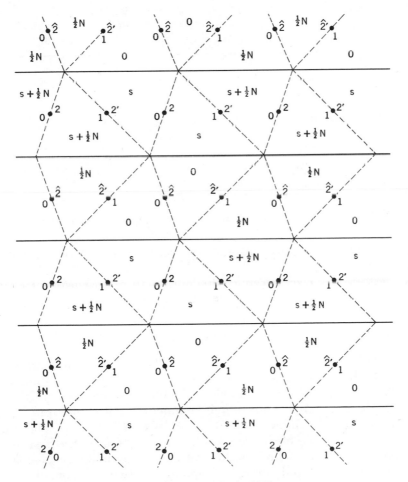

Fig. 99b. $222'2'$ m/g, 0011.

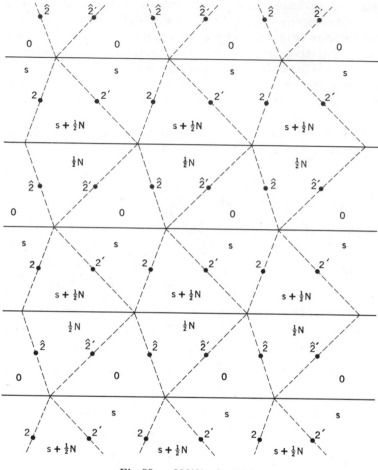

Fig. 99c.   222'2' m/g, 1111.

F.   Summary

The color configurations in the 22'2"2''' system are summarized in Table 14. It is noted that, except for a few degeneracies, this enumeration follows a regular pattern: when all color parameters vanish there is a single dichromatic and a single monochromatic configuration, the former corresponding to one already discussed in Chapters 8, 9, and 10, the latter to one of the geometric configurations of the first seven chapters. The other combinations of color parameters each have a single four-color configuration, and one or two dichromatic ones, depending on the degeneracy of the geometry.

Table 14 Colored configurations in the 22'2"2''' system

Rotocenters:

Symm. value	Color param.	Symm. value	Color param.	Symm. value	Color param.	Symm. value	Color param.	N	s
2	$\begin{cases}0\\0\\0\\1\end{cases}$	2'	$\begin{cases}0\\0\\1\\1\end{cases}$	2"	$\begin{cases}0\\1\\1\\1\end{cases}$	2'''	$\begin{cases}0\\1\\1\\1\end{cases}$	$\begin{cases}2\\1\end{cases}$ 4 2 $\begin{cases}4\\2\end{cases}$ $\begin{cases}4\\2\end{cases}$	$\begin{cases}1\\0\end{cases}$ 1 $\begin{cases}0\\1\end{cases}$ $\begin{cases}1\\0\end{cases}$ $\begin{cases}1\\0\end{cases}$
2	$\begin{cases}0\\0\\1\\1\end{cases}$	$\hat{2}$	$\begin{cases}0\\0\\1\\1\end{cases}$	2'	$\begin{cases}0\\1\\0\\1\end{cases}$	2"	$\begin{cases}0\\1\\0\\1\end{cases}$	$\begin{cases}2\\1\end{cases}$ $\begin{cases}4\\2\end{cases}$ 4 2 4 2	$\begin{cases}1\\0\end{cases}$ $\begin{cases}1\\0\end{cases}$ 1 $\begin{cases}0\\1\end{cases}$ 1 $\begin{cases}0\\1\end{cases}$
2	$\begin{cases}0\\0\\1\end{cases}$	$\hat{2}$	$\begin{cases}0\\0\\1\end{cases}$	2'	$\begin{cases}0\\1\\1\end{cases}$	$\hat{2}'(g/g')$	$\begin{cases}0\\1\\1\end{cases}$	$\begin{cases}2\\1\end{cases}$ $\begin{cases}4\\2\end{cases}$ $\begin{cases}4\\2\end{cases}$	$\begin{cases}1\\0\end{cases}$ $\begin{cases}1\\0\end{cases}$ $\begin{cases}1\\0\end{cases}$
2	$\begin{cases}0\\0\\1\end{cases}$	$\hat{2}$	$\begin{cases}0\\0\\1\end{cases}$	2'	$\begin{cases}0\\1\\1\end{cases}$	$\hat{2}'(m/g)$	$\begin{cases}0\\1\\1\end{cases}$	$\begin{cases}2\\1\end{cases}$ 4 2 4 2	$\begin{cases}1\\0\end{cases}$ 1 $\begin{cases}0\\1\end{cases}$ 1 $\begin{cases}0\\1\end{cases}$

22

Conclusions and summary

The generation of the last configuration in the 22'2"2''' system might have been the decisive finale of this monograph. The plane has been covered exhaustively, all impossible combinations of symmetry elements have been eliminated by a chain of theorems, and presumably all remaining configurations have been enumerated exhaustively.

Our approach has been algorismatic: patterns were generated from individual modules (motifs or symmetry elements) by logical interactions. Since in practice it is often necessary to identify symmetry elements in a complex pattern, it would seem fitting to conclude this monograph, which has been *synthetic* in its approach, with some *analytical* examples.

Among the most fascinating periodic patterns having color symmetry are the creations of M. C. Escher; many of these are represented in MacGillavry's "Symmetry aspects of M. C. Escher's periodic drawings" (Oosthoek). Because of the crystallographically unconventional approach of our monograph it is interesting to check it out against MacGillavry's own authoritative interpretations of the same drawings. The drawings selected are plates 37, 39, and 41 of the book by MacGillavry (Reproduced in black and white as Figs. 100, 101, and 102). None of these have enantiomorphy, so their fundamental regions can be chosen arbitrarily, and were chosen by Escher to have the shapes of a fish, a lizard, and apparently an unsymmetrical moth.

In Fig. 100 there is a fourfold rotocenter at the meeting point of four fishtails, and a second one, distinct from the first, where four fins meet. Four colors meet, in counterclockwise sequence white-brown-red-blue, at each fourfold center. We shall set white $= 0$, brown $= 1$,

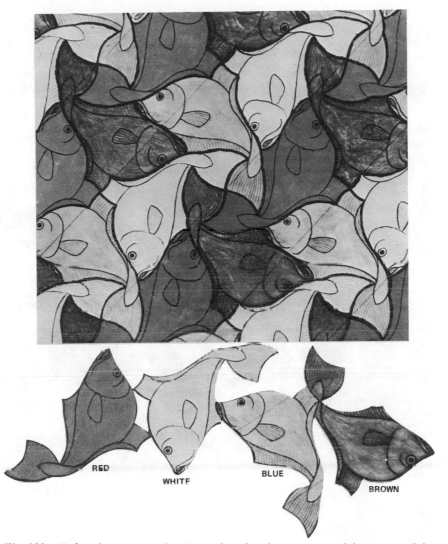

Fig. 100.   Fishes, by M. C. Escher (reproduced with permission of the artist and the
Escher Foundation).

red $= 2$, blue $= 3$; since these are four colors, $N = 4$. The pattern
belongs to the 244' system: the twofold rotocenters occur at the fishes'
jaws. Since four colors meet at the fourfold rotocenters in the same
sequence, $q = r = 1$; according to the second diophantine equation p
therefore equals unity. The color difference between fishes rubbing
jaws at twofold rotocenters equals $(p/k)N = \frac{1}{2} \times 4 = 2$; the drawing

Fig. 101.   Lizards, by M. C. Escher (reproduced with permission of the artist and the Escher Foundation).

164

does indeed show that white and red fishes rub jaws as do blue and brown ones, but never blue and red, white and brown, etc., for their colors do not differ by two units. All is therefore well.

Figure 101 is tricolored: $N = 3$. Twofold rotocenters occur at the right-hand wrists of the lizards, which touch at the tail tips of two other, equivalent, lizards. These twofold rotocenters relate lizards of identical colors: $k = 2$, $p = 0$. Three lizards of different colors meet at their right knees: the colors are consistently in counterclockwise sequence: black-red-white. We shall set black $\equiv 0$, red $\equiv 1$, white $\equiv 2$. The pattern is characterized by $l = 3$, and, since the colors around a threefold rotocenter are represented by the sequence $(0, \frac{1}{3}qN, \frac{2}{3}qN)$, $q = 1$. According to the first diophantine equation, sixfold rotocenters are implied: $m = 6$. These sixfold rotocenters occur at the lizards' left thumbs. According to the second diophantine equation $r = 4$. The color sequence around a sixfold rotocenter is $(0, \frac{1}{6}rN, \frac{1}{3}rN, \frac{1}{2}rN, \frac{2}{3}rN, \frac{5}{6}rN)$; with $r = 4$, $N = 3$, this becomes modulo-N: 0, 2, 1, 0, 2, 1, in other words, black-white-red-black-white-red. This is indeed the case, so again all is well.

In Fig. 102 MacGillavry considers each moth as one fundamental region; each moth has three colors, one for its body, one for its front wings, and one for its rear wings. Each front wing moreover contains a spot in the same color as its rear wings. We have so far considered only monochromatic fundamental regions, but have indicated that in the case of multicolored fundamental regions each color *combination* should be considered as a single color as long as consistency is maintained. In agreement with MacGillavry we could call this pattern a six-colored one. Since the body of each moth has just that color *not* shown on its wings, and since the spots on the front wings are the same color as the rear wings, each moth has its color completely specified by the colors of its front wings and their spots. A pair of symbols (r = red, b = brown, y = yellow) is used to designate each moth, the first indicating the color of the front wings, the second that of their spots. For example, yr indicates a moth having yellow front wings with red spots, red rear wings and a brown body. Figure 103*a* is a schematic interpretation of Fig. 102 in these terms.

Inspection of Fig. 103a reveals several startling characteristics. Consider first any twofold rotocenter: it is surrounded by four moths at the vertices of a rectangle. This rotocenter relates diagonally opposite moths to each other. It will be observed that the effect of the twofold rotocenter on one pair of moths is to interchange the colors of their front and rear wings; the same rotocenter retains the colors of the front wings of the other pair, changing only the colors of their rear

RED
W / Brown & Yellow Dots

YELLOW
W / Brown & Red Dots

BROWN
W / Red & Yellow Dots

Fig. 102. Moths, by M. C. Escher (reproduced with permission of the artist and the Escher Foundation).

166

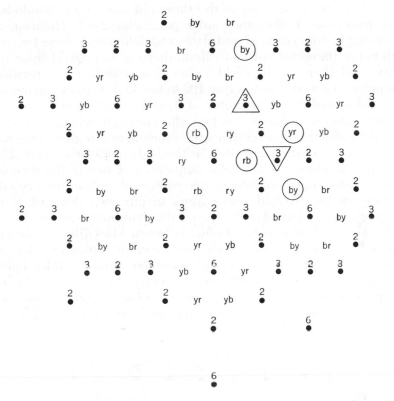

Fig. 103a.   Schematic representation of MacGillavry's interpretation of Fig. 102.

wings. Since the moths are presumably all equivalent because of the sixfold rotational symmetry, such different effects of the same symmetry element on two equivalent points is problematical. However, if we had followed MacGillavry's notation, using six letters instead of six pairs of three letters, there would be no problem here: the pattern could have been made up of moths of six different colors.

More paradoxical is the relation between two nearest threefold rotocenters, for example, those denoted by triangles in Fig. 103a. Because of the nearby sixfold rotocenter (or the twofold rotocenter between them) these two rotocenters are equivalent and should therefore have the same color parameter. If we look at the circled point *yr* in Fig. 103, however, we see that the *upper* threefold rotocenter relates it to equivalent points in the counterclockwise sequence *yr-by-rb*. The *lower* threefold rotocenter relates it to equivalent points in the counterclockwise color sequence *yr-rb-by*. Since these sequences are each other's opposites, one of the threefold rotocenters would have color parameter 1, the other color parameter 2 (cf. Theorem 27). According to our system of postulates and definitions, these two rotocenters are therefore *not* equivalent, there is *no* twofold color symmetry, and there is threefold rather than sixfold color rotational symmetry. We would assign Fig. 102 to the 33′3″ system, having only *three* colors and color parameters 0, 1, and 2 (Fig. 103b). The fundamental region would consist of two adjacent moths whose front wings have the color that characterizes the fundamental region. (One moth has rear wings of the same color as the body of the other moth.) The opposite directions of the color sequences of two of the threefold rotosimplexes represent information contained, in MacGillavry's description, in her twofold rotocenters; in our own description this information is contained in the second diophantine equation.

There is no fundamental conflict between MacGillavry's description and our own: each depends on its own definition of color rotocenters. If we interpret color as elevation out of the euclidean plane, a comparison can be made with the symmetry of a regular octahedron. The projection of this octahedron onto a plane parallel to one of its faces has sixfold rotational symmetry; analogously Fig. 102 has six-

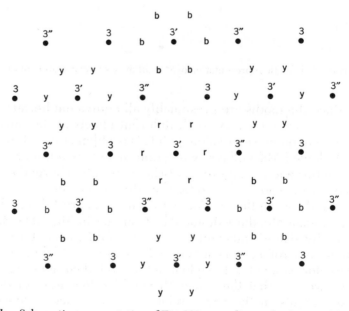

Fig. 103*b*. Schematic representation of Fig. 102, according to the theory of the present monograph.

fold geometrical symmetry if color differences are ignored. Nevertheless, the octahedron itself has no sixfold rotational symmetry: its symmetry around an axis perpendicular to one of its faces is only threefold. Conceivably, the octahedron could have a sixfold rotoreflection axis, the reflection being in a plane equidistant between opposite faces. However, such an axis is not in common use, the rotoinversion axis being used instead. Therefore in any system based on the rotoinversion axis rather than a rotoreflection axis, the regular octahedron has threefold, not sixfold rotational symmetry.

Similarly, MacGillavry's analysis of Escher's moth pattern is based on a definition of a sixfold color rotation center that differs from our own. Since MacGillavry's book does not aim at a self-consistent theory of color symmetry, no explicit definitions are given, although her analysis implies certain characteristics of the color symmetry elements used. We have found these characteristics to be generally in agreement with our own explicit definitions, except for the ambiguity in sixfold color rotocenters that exactly parallels the discrepancies between systems using rotoreflection axes and those using rotoinversion axes.

This final example serves to illustrate what has been said before: definitions of color and color symmetry are to a certain extent arbitrary, and exhaustiveness and classifications depend on these definitions. One set of definitions may be as valid as another; the value of a theory depends on the self-sufficiency of its definitions, and the internal consistency of the theory. It appears that the present theory uses definitions that do not conflict with generally accepted ideas about color symmetry. These definitions were arrived at in the form here presented by an inductive process, during which more and more internal inconsistencies were removed until, hopefully, none remained.

Bibliography

E. Alexander and K. Hermann, Die 80 zweidimensionaler Raumgruppen, Z. f. Krist., **70** (1929), 328–345.

G. Donnay, N. V. Belov, N. N. Neronova, and T. S. Smirnova, The Shubnikov groups, Sov. Phys. Cryst., **3** (1958), 642–644.

G. Donnay, L. M. Corliss, J. D. H. Donnay, N. Elliott, and J. M. Hastings, Symmetry of magnetic structures; magnetic structure of chalcopyrite, Phys. Rev., **112** (1958), 1917–1923.

J. D. H. Donnay and G. Donnay, Tables de Groupes Spatiaux Magnetiques, Compt. rend. Acad. Sci. Paris, **248** (1959), 3317–3319.

E. S. Federov, Zapiski Miner. Imper. S. Petersburg Obshch., **28** (1891), 345–390 and tables.

R. Fricke and F. Klein, Vorles, ue. d. Theorie der Automorphen Functionen. Bd. I. Leipzig: Teubner (1897).

H. Heesch, Zur Struktur theorie der ebener Symmetriegruppen, Z. f. Krist., **71** (1930), 95–102.

W. T. Holser, Relation of symmetry to structure in twinning. Z. f. Krist., **110** (1958), 249–265.

W. T. Holser, Point groups and plane groups in a two-sided plane and their subgroups, Z. F. Krist., **110** (1958), 266–281.

W. T. Holser, Classification of symmetry groups, Acta Cryst., **14** (1961), 1236–1242.

C. Jordan, Annali di Matematica da Brioschi e Cremona, Ser. II, T II (1869).

F. Klein and R. Fricke, Vorles, ue. d. Theorie d. Ellipt. Modulfunctionen, Bd. I. Leipzig: Teubner (1890).

V. A. Koptsik, Shubnikovskie Gruppy. Izdatal'stvo Moskovskogo Universiteta (1966).

G. Polya, Z. f. Krist., **60** (1924), 278–282.

A. Schoenflies, Theorie der Kristallstruktur. Berlin: Gebr. Borntraeger (1891; second edition, 1923).

171

L. Sohncke, *Borchard's Journal fuer reine und angew. Mathematik*, **77** (1874), 47–102.

A. Speiser, *Die Theorie der Gruppen von endlicher Ordnung*, Ch. VI. Berlin: Julius Springer Verlag (1922; reprint of third edition by Dover Publ., New York, 1943).

F. Steiger, *Comm. Math. Helv.*, **8** (1936), 235–249.

L. Weber, Die Symmetrie homogener ebener Punktsysteme, *Z. f. Krist.*, **70** (1929), 309–327.

Theorems

Theorem 1. The coexistence in a plane of two rotocenters implies the existence of a third rotocenter in the same plane. The symmetry values k, l, m of the triplet of rotocenters so related are constrained by the equation

$$\frac{1}{k}+\frac{1}{l}+\frac{1}{m}=1.$$

Theorem 2. The coexistence in a plane of two noncongruent twofold rotocenters precludes the existence in that plane of any rotocenters whose symmetry value does not equal *two*.

Theorem 3. The coexistence in a plane of two noncongruent threefold rotocenters implies the existence in that plane of three threefold rotosimplexes. Any straight line joining two noncongruent threefold rotocenters must pass through a threefold rotocenter that is not congruent with either of these two rotocenters.

Theorem 4. The existence in a plane, periodic pattern of a fourfold rotocenter implies the existence of two fourfold rotosimplexes as well as a twofold rotosimplex.

Theorem 5. The existence in a plane, periodic pattern of a sixfold rotocenter implies the existence of a twofold, a threefold, and a sixfold rotosimplex.

Theorem 6. In a plane, the rotosimplex having the highest finite symmetry value constitutes a lattice.

Theorem 7. When k-fold, l-fold, and m-fold rotocenters coexist in a plane, $k \leqslant l \leqslant m$ and m finite, the k-fold rotocenters have their environments oriented in m/k different directions, and l-fold rotocenters have their environments oriented in m/l different directions.

Theorem 8. Two sixfold rotocenters in a plane necessarily belong to the same lattice.

Theorem 9. A threefold and a fourfold rotocenter cannot coexist in a plane.

Theorem 10. The existence of a fivefold rotocenter in a plane precludes the existence of any other rotocenter in the same plane.

Theorem 11. The existence in a plane of a rotocenter with a finite symmetry value greater than six precludes the existence of any other rotocenter in that plane.

Theorem 12. The point bisecting a line segment joining two evenfold rotocenters belonging to the same lattice coincides with an evenfold rotocenter.

Theorem 13. Two points that are both congruent and enantiomorphic necessarily lie on mirror lines.

173

Theorem 14. In a pattern that possesses enantiomorphy every rotocenter that has no noncongruent enantiomorph necessarily lies on a mirror line. In particular, any sixfold rotocenter in a pattern that possesses enantiomorphy must lie on a mirror line.

Theorem 15. In a pattern that possesses enantiomorphy, every rotocenter not located on a mirror line is limited to the symmetry values 2, 3, 4, and ∞ and possesses a noncongruent enantiomorph.

Theorem 16. Two enantiomorphic k-fold rotocenters imply the intersection, at a point halfway between them, of either k glide lines or $(k-1)$ glide lines and a single mirror line.

Theorem 17. The acute angle between any two glide lines in a plane is limited to the values 0, 45, 60, and 90°. The coexistence of such glide lines implies the existence of rotosimplexes whose symmetry numbers have the respective combinations $1 \infty \infty, 244, 333, 2222$.

Theorem 18. The acute angle between a mirror line and a glide line in a plane is limited to the values 0, 45, 60, and 90°. The coexistence of such reflection lines implies the existence of rotosimplexes whose symmetry numbers have the respective combinations $1 \infty \infty, 244, 333, 22 \infty$. The line segment joining enantiomorphically paired rotocenters is perpendicularly bisected by the mirror line at the point of intersection of the mirror and glide lines.

Theorem 19. If two mirror lines intersect at an angle $\theta = j\pi/k$ where j and k are relatively prime integers, $2j \leq k$, then a k-fold center of rotational symmetry is implied at the intersection of the mirrors. There are k mirrors intersecting at the center of symmetry. If k is odd, these mirrors are all equivalent, but if k is even, there are two distinct sets of mirrors, each containing $\frac{1}{2}k$ mirrors.

Theorem 20. A rotocenter located on a glide line is limited to the symmetry numbers 2, 4, and ∞ and implies its enantiomorph on the same glide line. The shortest distance between enantiomorphs equals the translation component of the glide line.

Theorem 21. A k-fold rotocenter on a mirror line implies k mirror lines intersecting at equal angles at the center. If k is odd, all mirror lines are equivalent, but if k is even, the mirrors belong to two distinct sets.

Theorem 22. The coexistence in a plane of two equivalent, parallel mirror lines implies a mirror line parallel to and equidistant from these two mirros.

Theorem 23. The existence of a mirror line perpendicular to the direction of translational symmetry implies a set of equidistant, mutually parallel, alternatingly distinct mirror lines.

Theorem 24. Reversing rotocenters necessarily have even symmetry numbers; only evenfold rotocenters may be reversing.

Theorem 25. If the coexistence of two symmetry elements implies a third symmetry element, the number of reversing elements in this triplet must be even (either zero or 2).

Theorem 26. Of four coexisting twofold rotosimplexes the number of color-reversing ones must be even (0, 2, or 4).

Theorem 26a. When four rotosimplexes coexist in a plane, the sum of their color parameters must be even.

Theorem 27. Two colored configurations about different k-fold rotocenters whose color parameters p and p' are related by the expression

$$p' \overset{k}{=} k - p$$

correspond to the same color sequence being read in opposite directions.

Theorem 28. The color parameters p, q, and r of three coexisting rotocenters in a plane either all vanish or are related by the diophantine equation

$$\frac{p}{k} + \frac{q}{l} + \frac{r}{m} = 1,$$

where k, l, and m are the respective symmetry values of these rotocenters.

Theorem 29. The existence of a single pair of isochromatic enantiomorphs in a pattern implies that any point in the pattern equivalent to this pair has an isochromatic enantiomorph somewhere in the pattern.

Fig. 88. Exhaustive   illustration   of   color   configurations © Kennecott   Copper
Corporation.

2 2' ∞, 00; N=2, s=1

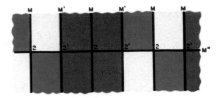

2 2' ∞, 01, N=4, s=1

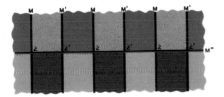

2 2' ∞, 01, N=2, s=0

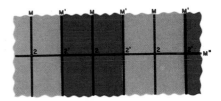

2 2' ∞, 01, N=2, s=1

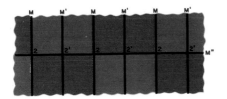

2 2' ∞, 11, N=4, s=1

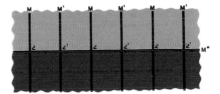

2 2' ∞, 11, N=2, s=0

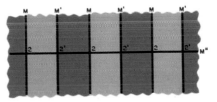

2 2' ∞, 11, N=2, s=1

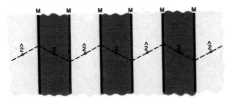

2 2̂ ∞, 00, N=2, s=1

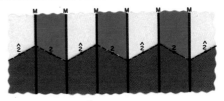

2 2̂ ∞, 11, N=4, s=1

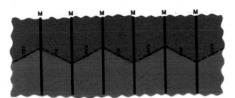

2 2̂ ∞, 11, N=2, s=0

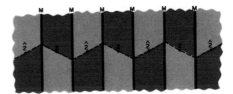

2 2̂ ∞, 11, N=2, s=1

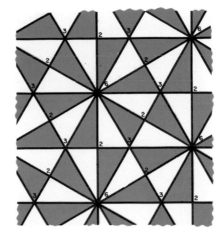

$\underline{2}\underline{3}\underline{6}$, 000, N=2, s=1

$\underline{2}\underline{3}\underline{6}$, 014, N=6, s=1

$\underline{2}\underline{3}\underline{6}$, 014, N=3, s=0

$\underline{2}\underline{3}\underline{6}$, 014, N=3, s=1

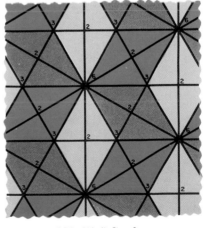

$\underline{2}\underline{3}\underline{6}$, 014, N=3, s=2

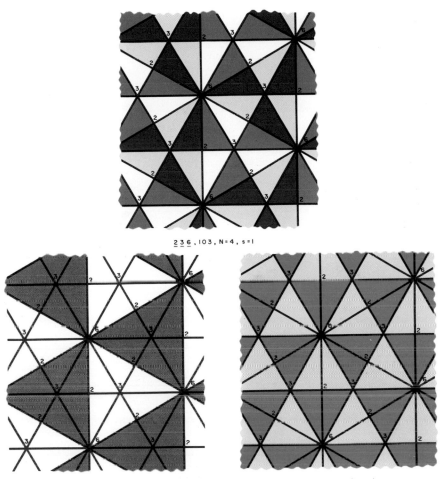

$\underline{2}\,\underline{3}\,6$, 103 , N=4 , s=1

$\underline{2}\,3\,\underline{6}$, 103 , N=2 , s=0

$\underline{2}\,\underline{3}\,\underline{6}$, 103 , N=2 , s=1

$\underline{2}\,\underline{3}\,\underline{6}$, III , N=12 , s=1

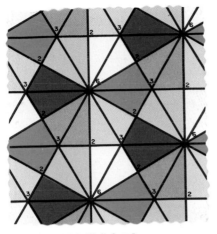

$\underline{2}\,\underline{3}\,\underline{6}$, III, N = 6, s = 0

$\underline{2}\,\underline{3}\,\underline{6}$, III, N = 6, s = 1

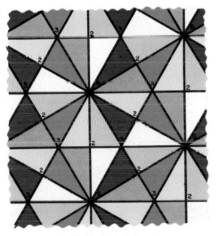

$\underline{2}\,\underline{3}\,\underline{6}$, III, N = 6, s = 2

236, III, N = 6, s = 3

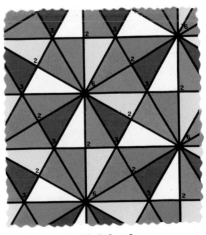

$\underline{2}\,\underline{3}\,\underline{6}$, III, N = 6, s = 4

$\underline{2}\,\underline{3}\,\underline{6}$, III, N = 6, s = 5

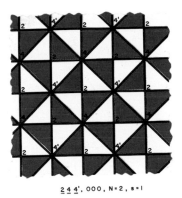

2 4 4', 000, N=2, s=1

2 4 4', 013, N=8, s=1

2 4 4', 013, N=4, s=0

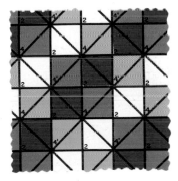

2 4 4', 013, N=4, s=1

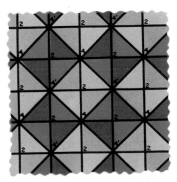

2 4 4', 013, N=4, s=2

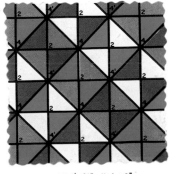

2 4 4', 013, N=4, s=3

$\underline{2}\,\underline{4}\,\underline{4}'$, 022, N=4, s=1

$\underline{2}\,\underline{4}\,\underline{4}'$, 022, N=2, s=0

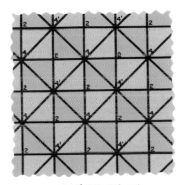

$\underline{2}\,\underline{4}\,\underline{4}'$, 022, N=2, s=1

$\underline{2}\,\underline{4}\,\underline{4}'$, 102, N=4, s=1

$\underline{2}\,\underline{4}\,\underline{4}'$, 102, N=2, s=0

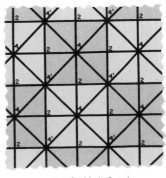

$\underline{2}\,\underline{4}\,\underline{4}'$, 102, N=2, s=1

2 4 4', III, N = 8, s = 1

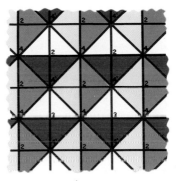

2 4 4', III, N = 4, s = 0

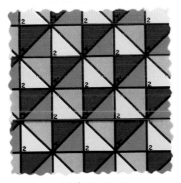

2 4 4', III, N = 4, s = 1

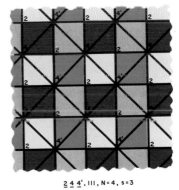

2 4 4', III, N = 4, s = 3

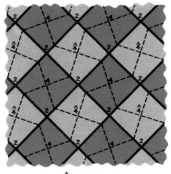

$\underline{2}\,4\,\hat{4},\,000,\,N=2,\,s=1$

$\underline{2}\,4\,\hat{4},\,013,\,N=8,\,s=1$

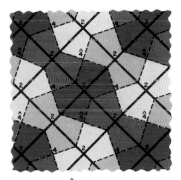

$\underline{2}\,4\,\hat{4},\,013,\,N=4,\,s=0$

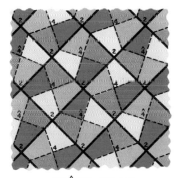

$\underline{2}\,4\,\hat{4},\,013,\,N=4,\,s=2$

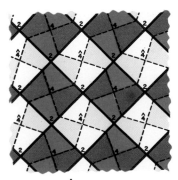

$\underline{2}\,4\,\hat{4},\,022,\,N=4,\,s=1$

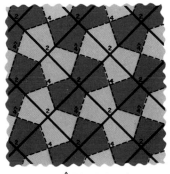

$\underline{2}\,4\,\hat{4},\,022,\,N=2,\,s=0$

$\underline{2}\ 4\ \hat{4},\ 022,\ N=2,\ s=1$

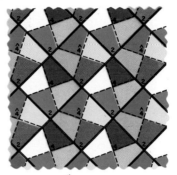

$\underline{2}\ 4\ \hat{4},\ III,\ N=8,\ s=1$

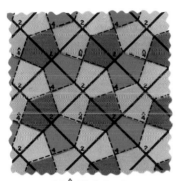

$\underline{2}\ 4\ \hat{4},\ III,\ N=4,\ s=0$

$\underline{2}\ 4\ \hat{4},\ III,\ N=4,\ s=1$

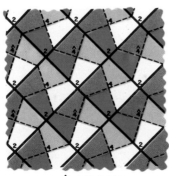

$\underline{2}\ 4\ \hat{4},\ III,\ N=4,\ s=2$

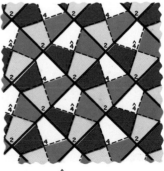

$\underline{2}\ 4\ \hat{4},\ III,\ N=4,\ s=3$

$\underline{3}\,\underline{3}'\underline{3}'',\ 000,\ N{=}2,\ s{=}1$

$\underline{3}\,\underline{3}'\underline{3}'',\ 012,\ N{=}6,\ s{=}1$

$\underline{3}\,\underline{3}'\underline{3}'',\ 012,\ N{=}3,\ s{=}0$

$\underline{3}\,\underline{3}'\underline{3}'',\ 012,\ N{=}3,\ s{=}1$

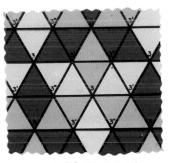

$\underline{3}\,\underline{3}'\underline{3}'',\ 012,\ N{=}3,\ s{=}2$

$\underline{3}\,\underline{3}'\underline{3}'',\ 111,\ N{=}6,\ s{=}1$

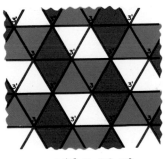

$\underline{3}\,\underline{3}'\underline{3}'',\ 111,\ N{=}3,\ s{=}0$

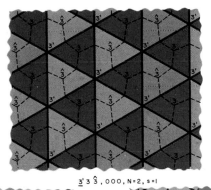

$\underline{3}'\,3\,\hat{3}$, 000, N=2, s=1

$\underline{3}'\,3\,\hat{3}$, 012, N=6, s=1

$\underline{3}'\,3\,\hat{3}$, 012, N=3, s=0

$\underline{3}'\,3\,\hat{3}$, III, N=6, s=1

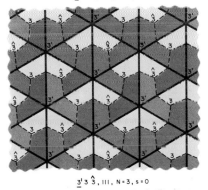

$\underline{3}'\,3\,\hat{3}$, III, N=3, s=0

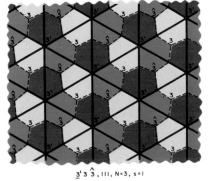

$\underline{3}'\,3\,\hat{3}$, III, N=3, s=1

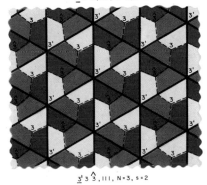

$\underline{3}'\,3\,\hat{3}$, III, N=3, s=2

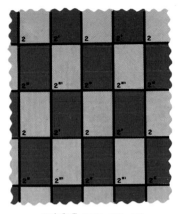

$\underline{2}\,\underline{2}'\underline{2}''\underline{2}'''$, 0000, N=2, s=1

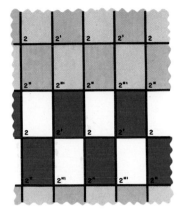

$\underline{2}\,\underline{2}'\underline{2}''\underline{2}'''$, 0011, N=4, s=1

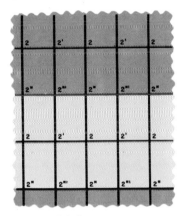

$\underline{2}\,\underline{2}'\underline{2}''\underline{2}'''$, 0011, N=2, s=0

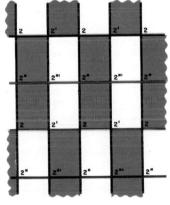

$\underline{2}\,\underline{2}'\underline{2}''\underline{2}'''$, 0011, N=2, s=1

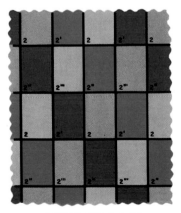

$\underline{2}\,\underline{2}'\underline{2}''\underline{2}'''$, 0110, N=4, s=1

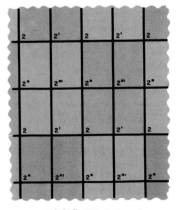

$\underline{2}\,\underline{2}'\underline{2}''\underline{2}'''$, 0110, N=2, s=0

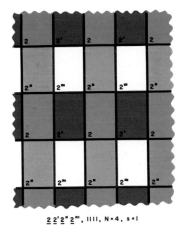

$\underline{2}\,\underline{2}'\underline{2}''\underline{2}'''$, IIII, N=4, s=I

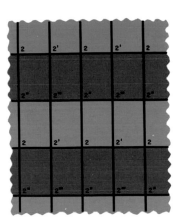

$\underline{2}\,\underline{2}'\underline{2}''\underline{2}'''$, IIII, N=2, s=0

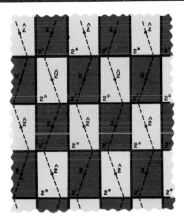

$2\,\overset{\wedge}{2}\,\underline{2}'\underline{2}''$, 0000, N=2, s=I

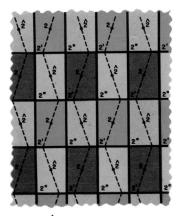

$2\,\overset{\wedge}{2}\,\underline{2}'\underline{2}''$, 00II, N=4, s=I

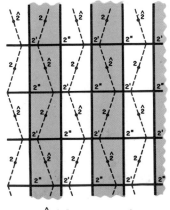

$2\,\overset{\wedge}{2}\,\underline{2}'\underline{2}''$, 00II, N=2, s=0

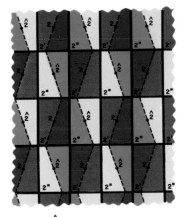

$2 \hat{\underline{2}} \underline{2}' \underline{2}''$, 1100, N = 4, s = 1

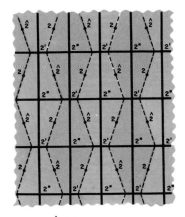

$2 \hat{\underline{2}} \underline{2}' \underline{2}''$, 1100, N = 2, s = 0

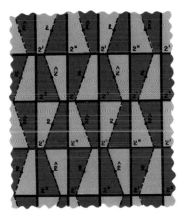

$2 \hat{\underline{2}} \underline{2}' \underline{2}''$, 1100, N = 2, s = 1

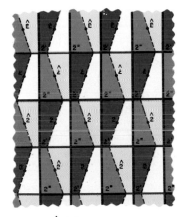

$2 \hat{\underline{2}} \underline{2}' \underline{2}''$, 1111, N = 4, s = 1

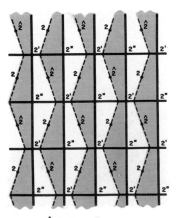

$2 \hat{\underline{2}} \underline{2}' \underline{2}''$, 1111, N = 2, s = 0

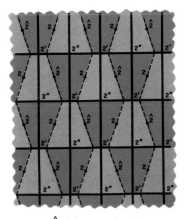

$2 \hat{\underline{2}} \underline{2}' \underline{2}''$, 1111, N = 2, s = 1

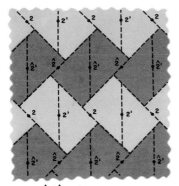

$2 \overset{\wedge}{2} 2'\overset{\wedge}{2}'$ g/g', 0000, N=2, s=1

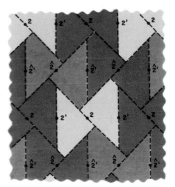

$2 \overset{\wedge}{2} 2'\overset{\wedge}{2}'$ g/g', 0011, N=4, s=1

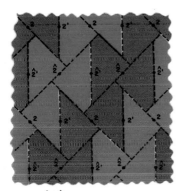

$2 \overset{\wedge}{2} 2'\overset{\wedge}{2}'$ g/g', 0011, N=2, s=0

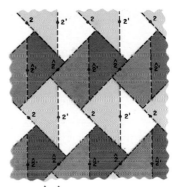

$2 \overset{\wedge}{2} 2'\overset{\wedge}{2}'$ g/g', 1111, N=4, s=1

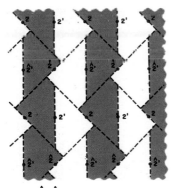

$2 \overset{\wedge}{2} 2'\overset{\wedge}{2}'$ g/g', 1111, N=2, s=0

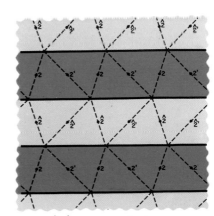

$2 \overset{\wedge}{2} 2'\overset{\wedge}{2}'$ m/g, 0000, N=2, s=1

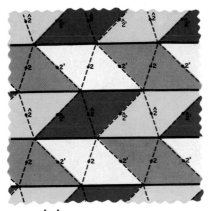

$2 \overset{\wedge}{2} 2' \overset{\wedge}{2}' \, m/g, \, 0011, \, N=4, \, s=1$

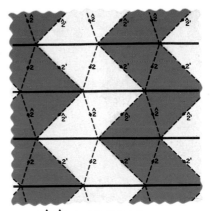

$2 \overset{\wedge}{2} 2' \overset{\wedge}{2}' \, m/g, \, 0011, \, N=2, \, s=0$

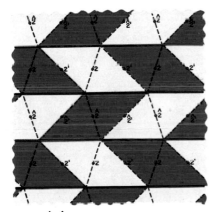

$2 \overset{\wedge}{2} 2' \overset{\wedge}{2}' \, m/g, \, 0011, \, N=2, \, s=1$

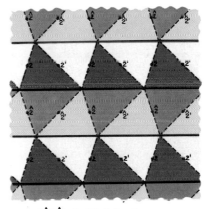

$2 \overset{\wedge}{2} 2' \overset{\wedge}{2}' \, m/g, \, 1111, \, N=4, \, s=1$

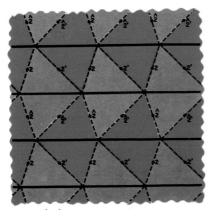

$2 \overset{\wedge}{2} 2' \overset{\wedge}{2}' \, m/g, \, 1111, \, N=2, \, s=0$

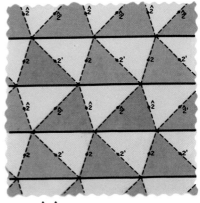

$2 \overset{\wedge}{2} 2' \overset{\wedge}{2}' \, m/g, \, 1111, \, N=2, \, s=1$

Index

Items that recur throughout this monograph are listed only at their initial appearance, definition, or in another special context.